T0142481

Springer Theses

Recognizing Outstanding Ph.D. Research

Aims and Scope

The series "Springer Theses" brings together a selection of the very best Ph.D. theses from around the world and across the physical sciences. Nominated and endorsed by two recognized specialists, each published volume has been selected for its scientific excellence and the high impact of its contents for the pertinent field of research. For greater accessibility to non-specialists, the published versions include an extended introduction, as well as a foreword by the student's supervisor explaining the special relevance of the work for the field. As a whole, the series will provide a valuable resource both for newcomers to the research fields described, and for other scientists seeking detailed background information on special questions. Finally, it provides an accredited documentation of the valuable contributions made by today's younger generation of scientists.

Theses are accepted into the series by invited nomination only and must fulfill all of the following criteria

- They must be written in good English.
- The topic should fall within the confines of Chemistry, Physics, Earth Sciences, Engineering and related interdisciplinary fields such as Materials, Nanoscience, Chemical Engineering, Complex Systems and Biophysics.
- The work reported in the thesis must represent a significant scientific advance.
- If the thesis includes previously published material, permission to reproduce this must be gained from the respective copyright holder.
- They must have been examined and passed during the 12 months prior to nomination.
- Each thesis should include a foreword by the supervisor outlining the significance of its content.
- The theses should have a clearly defined structure including an introduction accessible to scientists not expert in that particular field.

More information about this series at http://www.springer.com/series/8790

Ritika Dusad

Magnetic Monopole Noise

Doctoral Thesis accepted by Cornell
University, USA

 Springer

Ritika Dusad
Nucleus Software
New Delhi, India

ISSN 2190-5053   ISSN 2190-5061   (electronic)
Springer Theses
ISBN 978-3-030-58195-4   ISBN 978-3-030-58193-0   (eBook)
https://doi.org/10.1007/978-3-030-58193-0

This Springer imprint is published by the registered company Springer Nature Switzerland AG
The registered company address is: Gewerbestrasse 11, 6330 Cham, Switzerland

This thesis is dedicated to my dearest and most beloved mother. Her constant support has emboldened me to travel on this challenging but very beautiful journey.

Supervisor's Foreword

For her Ph.D. research, Ritika Dusad invented an ultra-sensitive, spin noise spectrometer and used it to discover the magnetic noise signature of emergent magnetic monopoles in $Dy_2Ti_2O_7$. Her vision for how to search for the monopoles was highly innovative. The elegance and simplicity of the design of her superconducting quantum interference device based spectrometer is striking and continues to inform the design of next generation experiments. The sensitivity of all her experimental measurements was at the limit of physical possibility and deeply impressive. Her realization of the correct microscopic theory underpinning her discoveries was inspiring. Ritika Dusad has that rare combination of wide scientific curiosity, sharp focus on objectives, brilliant intuition, and agile execution of her research plans. Perhaps even most impressive is her independence, her supply of new ideas, her enthusiasm for new challenges, and her delight in discovery. It was a highly rewarding experience and a great privilege to have Ritika Dusad as a colleague, and I very much look forward to renewing these experiences in future research projects.

Cornell University, Ithaca, NY, USA   Prof. J.C. Sèamus Davis

Abstract

Magnetic monopoles are hypothetical elementary particles exhibiting quantized magnetic charge $m_0 = \pm(h/(\mu_0 e))$ and quantized magnetic flux $\Phi_0 = \pm h/e$. In principle, such a magnetic charge can be detected by the quantized jump in magnetic flux Φ it generates upon passage through a superconducting quantum interference device (SQUID). Naturally, with the theoretical discovery that a plasma of emergent magnetic charges should exist in several lanthanide-pyrochlore magnetic insulators, including $Dy_2Ti_2O_7$, this SQUID technique was proposed for their direct detection. Experimentally, this has proven challenging because of the high number density of the monopole plasma. Recently, however, theoretical advances have allowed the spectral density of magnetic-flux noise $S_\Phi(\omega, T)$ due to generation recombination fluctuations of $\pm m_*$ magnetic charge pairs to be predicted. Here we report development of a SQUID based flux-noise spectrometer and consequent measurements of the frequency and temperature dependence of $S_\Phi(\omega, T)$ for $Dy_2Ti_2O_7$ samples. Virtually all the elements of $S_\Phi(\omega, T)$ predicted for a magnetic monopole plasma, including the existence of intense magnetization noise and its characteristic frequency and temperature dependence, are detected. Moreover, comparisons of simulated and measured correlation functions $C_\Phi(t)$ of the magnetic-flux noise $\Phi(t)$ imply that the motion of magnetic charges is strongly correlated.

Acknowledgements

I thank my advisor, Prof. J.C. Sèamus Davis for being a source of guidance, inspiration, and support. His brilliant ideas and deep knowledge of Physics never cease to amaze and inspire me. He has always encouraged me to work on fundamental problems while paying careful attention to details. His advice on how to achieve career goals has often been very useful and I cannot imagine an advisor more supportive than him on such issues. It has been very stimulating and rewarding to collaborate with him on my thesis project. I thank Prof. Steve J. Blundell and Prof. Graeme Luke for an exciting and rewarding collaboration. I thank my committee members, Prof. Katja Nowack, Prof. Michael Lawler, and Prof. J.C. Sèamus Davis for providing very useful feedback while I was working on my thesis.

My lab mates Yi Xue Chong, Stephen Edkins, Peter Sprau, Rahul Sharma, and Andrey Kostin have made my time in the Davis group a very joyful one. Our lunches, dinners, and social outings made life quite spirited. Anna Eyal and Azar Eyvazov taught me innumerable things and helped me transition into experimental physics seamlessly. I thank Yi Xue in particular for being one of my best friends in Ithaca and helping me out with my cryostat operations innumerable times. Were it not for Yi Xue's help, I would not be able to do the experiments onboard the 1K cryostat presented in this thesis. Steve has been unimaginably generous with his time in terms of helping me out with postdoc applications or any other major discussions about scientific careers. I have learned a lot about tact and professionalism from him. Peter is one of the most wise men I have met in my life—his words of wisdom have helped me through some of the trickiest situations I have faced. Rahul and Andrey are among the smartest people I have met, and they inspire me to be a lot more careful with physics than I am. I thank Jesse Hoke, Ben R. Roberts, Elizabeth Donoway, and Andrew Parmet for our collaborations together.

I thank Jyoti Pandey, Pankaj Singh, Gauri Patwardhan, Aniket Kakatkar, Chaitanya Joshi, Pooja Gudibanda, Ayush Dubey, Sharvil Talati, Arzoo Katiyar, Prateek Sehgal, Ayan Bhattacharya, and Shachi Deshpande for entertaining my enthusiasm for physics while we chatted about life over countless brunches and chai sessions. Ashudeep Singh took the beautiful photos of my cryostat presented in this thesis. I thank Dr. Gil Travish for mentoring me for close to ten years in matters of science and life. I thank Prof. Sanjay Puri and Dr. T.D. Senguttuvan for their ever enthusiastic view of physics and academia that has always motivated me to someday go back to my home country. Jay Joshi and Manasi Patwardhan have been my guardian angels in this foreign land, I'm indebted to them for this. I thank all of my friends and family I have missed mentioning by name here.

I thank Kacey Acquilano, Douglas Milton, Judy Wilson, Robert Sprankle, and Jonathan Fuller for helping me traverse administrative nooks and corners in graduate school. I thank Nathan Ellis for teaching me very patiently how to machine parts out of a large variety of materials. I thank Jeffrey Koski, Chris Cowulich, Stanley McFall, and Robert Page for machining the 1K cryostat and the tiny MACOR parts which I doubt other machinists would be able to pull off with such ease and precision. I thank Eric N. Smith for his boundless enthusiasm for and endless knowledge of Low Temperature Physics which he would freely share with students of this subject.

I thank Shri Krishna for introducing me to the beauty of physics at crucial moments of my life and blessing me with the opportunity of working on experimental physics when I least expected it.

Finally, I thank my wonderful parents and sister for always being there for me. It is extremely difficult to put into words the immense gratitude I feel towards them. I will not be able to do justice to this task but here's a small attempt. My father was one of the key people in my life who instigated my interest in science and physics with his emphasis on understanding things deeply. My sister and I spent numerous hours drinking copious amounts of Gimme coffee and discussing matters both scientific and personal. My mother and I chatted on the phone at random hours of the day, whenever I needed to speak to a close friend. The unwavering support of my family has been a great blessing and a source of strength that I drew upon throughout my graduate school time. Their visits to Ithaca have grounded me and made me realize that I am never far from my home country.

Parts of This Thesis Have Been Published in the Following Journal Articles

R. Dusad, F.K.K. Kirschner, J.C. Hoke, B. Roberts, A. Eyal, F. Flicker, G.M. Luke, S.J. Blundell, J.C. Seamus Davis, Magnetic monopole noise. Nature **571**, 234–239 (2019)

Contents

Chapter 1
Introduction

It has long been known that a magnet cannot be broken down to its simplest constituents—the north and south poles. This is reflected in the asymmetry of Maxwell's equations with respect to electric and magnetic fields. While the electric monopole is very much present and is the foundational basis of modern electronics, the magnetic monopole charge is conspicuous by its absence. The search for the fundamental magnetic monopole [1] has proven to be quite elusive till date, but condensed matter physics has provided a few candidates [2, 3]. This chapter reviews one such class of candidate materials, namely the Dipolar Spin Ices: Dysprosium Titanate and Holmium Titanate.

Lanthanide pyrochlore oxides are a class of materials wherein the rare-earth (RE) ions resides on corners of a tetrahedral network (Fig. 1.1) and the RE spins are usually frustrated. Magnetically frustrated systems [4] refer to conflict within different couplings between spins in a lattice. A simple illustration of frustration is a triangular lattice with Ising spins coupled antiferromagnetically. The attribute of frustration leads itself to the existence of a multitude of exotic magnetic states such as spin ices [5], spin slush [6] and candidates for quantum spin liquids [7]. One such member of this class presents itself as a candidate for housing magnetic monopoles, namely dipolar spin ice which is a geometrically frustrated magnet. Dysprosium Titanate ($Dy_2Ti_2O_7$) and Holmium Titanate ($Ho_2Ti_2O_7$) both belong to this sub-class of dipolar Spin Ices. Theorists predict that elementary spin excitations in these compounds behave like magnetic charges [8] that are deconfined to move about the lattice freely [3].

1.1 Spin Ices

The structure of $(Dy/Ho)_2Ti_2O_7$ is made up of a cubic unit cell (lattice parameter 10.12Å) with two sublattices—pyrochlore lattice of Dy ions (Fig. 1.1) with O at

© Springer Nature Switzerland AG 2021

R. Dusad, *Magnetic Monopole Noise*, Springer Theses,

https://doi.org/10.1007/978-3-030-58193-0_1

Fig. 1.1 Dy^{3+} ions represented by the small spheres sitting on a pyrochlore lattice structure

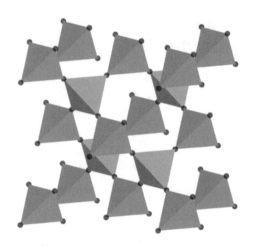

center of each tetrahedron and corner sharing octahedral network of Ti and O ions. The nearest neighbor distance between two Dy atoms is $a = 3.58$Å. The magnetic character of these materials arises from the lanthanide RE ions since the Ti and O ions are non-magnetic. Neutron scattering experiments first suggested that $Ho_2Ti_2O_7$ does not magnetically order down to 0.35K [9] in zero field, despite being chemically ordered. This strongly indicated frustration among Ho spins. When a sufficiently strong field was applied ($\sim$1 Tesla) appearance of Bragg peaks illustrated the restoration of magnetic order in this compound. In the conclusions of their study, Harris et al. recommended that the spin structure of this compound was consistent with an Ising like behavior of Ho^{3+} spins. This implied that Ho spins could only point towards or away from the center of the tetrahedron, this condition is equivalent to Ho spins pointing along local $\langle 111 \rangle$ axes.

The lowest energy state for this system corresponds to a 2-in-2-out spin configuration on each tetrahedron (Fig. 1.2b). Such a spin configuration is similar to proton arrangement in water ice (Fig. 1.2a). In H_2O ice, an oxygen atom sits at the center of a tetrahedron and two hydrogen atoms are placed relatively close to this atom with two other hydrogen atoms being placed relatively further away from the Oxygen. These rules of hydrogen arrangement around oxygen in ice are called Bernal-Fowler ice rules. It was found that Ho^{3+} ion ($4f^{10}$) could be replaced with Dy^{3+} ion ($4f^9$) while maintaining the ice rules. The similarity between structure of water ice and the spin states in $Dy/Ho_2Ti_2O_7$ allowed scientists to label these materials with a moniker 'Spin Ice'. A total of six spin arrangements that follow the ice rules are allowed for each tetrahedron in spin ice (out of a total 16 possible), making the overall crystal of Dy/Ho spins highly degenerate.

Calirometric measurements of $Dy_2Ti_2O_7$ in zero field revealed a broad peak in its specific heat $\sim$1K [10]. The lack of an ordering feature (sharp peak) in the specific heat $C(T)$ in the absence of a magnetic field (Fig. 1.3 top left) was consistent with the understanding of geometrical frustration present in spin ice [9]. The experimental technique used for the measurement of $C(T)$ involved application

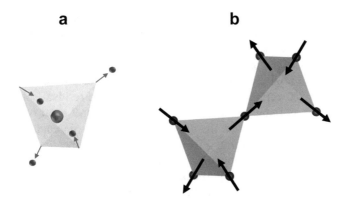

Fig. 1.2 (**a**) Structure of water ice consists of an oxygen atom (blue) at the center of a tetrahedron with two hydrogens (grey) bound close to the oxygen and two hydrogens sitting away from the oxygen, (**b**) Spin Ice configuration of $Dy_2Ti_2O_7$ where two Dy spins point into the center of the tetrahedron and two Dy spins point out; this is similar to structure of water ice

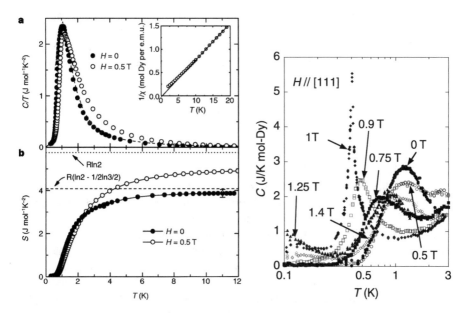

Fig. 1.3 Top left: Specific heat $C(T)$ measurements of $Dy_2Ti_2O_7$ show a lack of an ordering feature when the compound is in zero field. Bottom left: The residual spin entropy of $Dy_2Ti_2O_7$ is shown in this plot. Reprinted by permission from [Springer Nature Customer Service Centre GmbH]: [Nature] [10] (Zero-point entropy in 'spin ice', Ramirez et al.), [©] (1999). Right: Sharp peaks in the specific heat of $Dy_2Ti_2O_7$ appear when strong magnetic fields are applied. Figure reproduced with permission from from Ref. [11]. ©(2004) The Physical Society of Japan

of a semi-adiabatic heat pulse to the sample. The equilibration time window for the sample was set as 15 sec. Since the internal relaxation time of the sample merged with that of the technique below temperatures of ~ 0.2K, $C(T)$ was not well defined below that temperature. To test whether degeneracy of spin ice broke down at low temperatures, the total spin entropy of $Dy_2Ti_2O_7$ was determined by integrating $C(T)/T$ over a temperature range of 0.2K to 12K. The residual spin entropy of this compound was found to be $\Delta S = (0.67 \pm 0.04)R\ln 2$ short of $\sim 1/3$ from the expected spin entropy from the degeneracy of Ising spin configurations on a tetrahedron. In later experiments that waited for longer times (~ 1000s) for the system to equilibrate, this entropy was found to be restored to the system [12]. Such conflicting experimental results posed questions about the nature of spin ice ground state which remain to be resolved. Neutron time-of-flight measurements detailed how strong crystal fields in the spin ice compounds [13] resulted in splitting the degeneracy in f-shell occupancy of the RE^{3+} ions. The lowest energy state was found to be a doublet composed of $m_J = |\pm 8\rangle$ for $Ho_2Ti_2O_7$ and $|\pm 15/2\rangle$ for $Dy_2Ti_2O_7$.

The high-magnetic moment ($\mu \approx 10\mu_B$) of the rare-earth ions implicates long-range dipolar interactions[14] among the Dy spins. A Dipolar Spin Ice model (DSIM) was developed to describe the energetics of $Dy_2Ti_2O_7$ (DTO) and $Ho_2Ti_2O_7$ (HTO). This model (Eq. 1.1) includes both nearest neighbor exchange J and dipolar interactions D between the Dy/Ho spins to explain the spin correlations in these spin ices.

$$\mathcal{H} = -J\sum_{\langle i,j \rangle} \mathbf{S}_i \cdot \mathbf{S}_j + Da^3 \sum_{i<j} \left[\frac{\mathbf{S}_i \cdot \mathbf{S}_j}{|r_{ij}^3|} - \frac{3(\mathbf{S}_i \cdot r_{ij})(\mathbf{S}_j \cdot r_{ij})}{|r_{ij}|^5} \right] \qquad (1.1)$$

Here $\mathbf{S}_i^z$ is the Ising moment with magnitude of $|\mathbf{S}_i| = 1$ and D comes from a typical estimate of dipolar interaction energy $D = (\mu_0/4\pi)\mu^2/r_{nn}$, where r_{nn} is the nearest neighbor distance. J is a parameter determined from fits to measured specific heat [15]. For $Dy_2Ti_2O_7$, $J \approx 3.72$ K and the dipolar energy is $D \approx 1.41$ K.

DC susceptibility measurements of this compound implied a magnetic ordering temperature of $T_{CW} \approx 1.2K$ [16], where T_{CW} denotes Curie-Weiss temperature, i.e. x-axis intercept of inverse susceptibility vs T plot (Fig. 1.4). While in general long-range dipolar interactions between Dy spins are expected to lift the high degeneracy in spin ice states, evidence for such a magnetic ordering is not found for $Dy_2Ti_2O_7$ in its neutron scattering spectra. $Dy_2Ti_2O_7$ exhibits diffuse scattering (Fig. 1.5) at different temperatures—20K, 1.3K, 0.3K and 0.05K [17], i.e. both above and below the Curie-Weiss temperature for this compound. An absence of magnetic bragg peaks in the spectra throughout the temperature range indicates that magnetic order does not set in around T_{CW} or below.

The high degeneracy in spin ice states was further explored by studying the magnetization in response to a magnetic field. Stark difference between magnetization of $Dy_2Ti_2O_7$ when it was field-cooled (FC) and Zero-Field Cooled (ZFC) indicated history-dependence in magnetic moment of the compound at temperatures lower

Fig. 1.4 Reciprocal
susceptibility χ vs
temperature allows the
prediction of a magnetic
ordering temperature T_{CW}.
Reprinted figure with
permission from
R. Higashinaka et al., Phys.
Rev. B., 65, 054410 (2002)
Copyright (2002) by the
American Physical Society
[16]

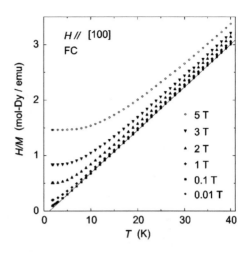

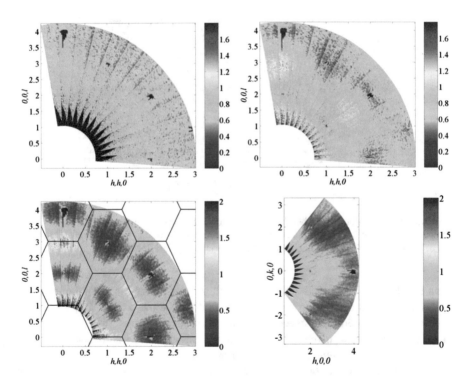

Fig. 1.5 $Dy_2Ti_2O_7$ neutron scattering spectra measured at a: 20K (top left), 1.3K (top right), 0.3K (bottom left) and 0.05K (bottom right). These experimentally measured spectra show diffuse scattering with no indication of magnetic ordering down to the lowest temperature. Reprinted figure with permission from T. Fennell et al., Phys. Rev. B., 73, 134408 (2004) Copyright (2004) by the American Physical Society [17]

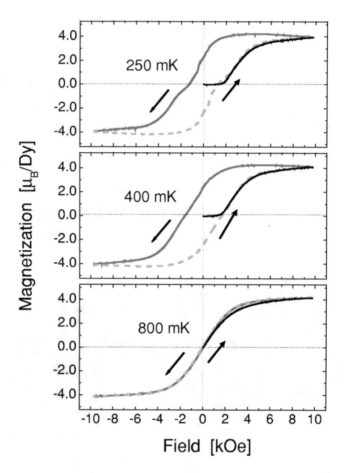

Fig. 1.6 Magnetization vs applied field curves for a $Dy_2Ti_2O_7$ sample exhibit hysteresis at temperatures below 0.65K. Reprinted figure with permission from J. Snyder et al., Phys. Rev. B., 69, 064414 (2004) Copyright (2004) by the American Physical Society [18]

than 0.65K [18]. Hysteresis was observed in the magnetization of this compound when a field was cycled through it at temperatures below 0.65K. This hysteresis vanished and a complete reversibility in magnetization was observed at 0.8K for $Dy_2Ti_2O_7$. These memory effects below a certain temperature indicated spin freezing of this compound below 0.65K (Fig. 1.6).

It is pertinent to note that the high degeneracy in spin ice states is lifted by the application of high enough magnetic fields $\sim 1T$ [19]. The signatures of this phase transition are visible both in the appearance of bragg peaks in neutron scattering spectra (Fig. 1.7) and onset of sharp peaks in specific heat vs temperature plots when a high magnetic field is applied [11].

AC susceptibility measurements (Fig. 1.8 left) allowed the determination of spin relaxation times for $Dy_2Ti_2O_7$ [21]. This revealed a few different regimes

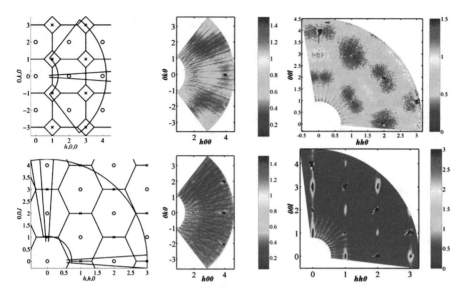

Fig. 1.7 $Dy_2Ti_2O_7$ neutron scattering spectra at 0.05K with zero field applied (top) and when a field of 1T is applied (bottom). Diffuse scattering disappears and magnetic bragg peaks appear in the presence of a magnetic field indicating the onset of magnetic order in this compound. Reprinted figure with permission from T. Fennell et al., Phys. Rev. B., 72, 224411 (2005) Copyright (2005) by the American Physical Society. Color version of scatter plots courtesy from Dr. O. Petrenko [20]

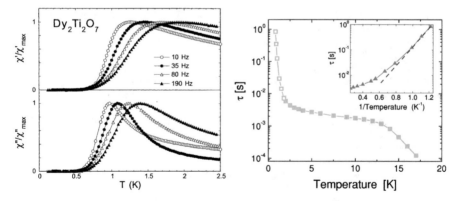

Fig. 1.8 left: AC susceptibility measurements of $Dy_2Ti_2O_7$ allow determination of spin relaxation times of this compound as a function of temperature. Used with permission from K Matsuhira et al Novel dynamical magnetic properties in the spin ice compound Dy2Ti2O7. J. Phys. Condens. Matter 13, L737 (2001) permission conveyed through Copyright Clearance Center, Inc. [21]. right: Spin relaxation times determined from susceptibility measurements of $Dy_2Ti_2O_7$ exhibit three different regimes including spin freezing at temperatures higher than 10K and lower than 4K. Reprinted figure with permission from J. Snyder et al., Phys. Rev. B., 69, 064414 (2004) Copyright (2004) by the American Physical Society [18]

in relaxation times (Fig. 1.8 right) of $Dy_2Ti_2O_7$ [22]: namely spin freezing at high temperatures around $\sim 16K$ as well as at low temperatures $T < 4K$ and an intermediate plateau region. Later studies of the spin relaxation in $Dy_2Ti_2O_7$ and $Ho_2Ti_2O_7$ revealed the existence of a supercooled spin liquid state in these compounds [23, 24] at lower temperatures. The exact microscopic mechanisms involved in the spin dynamics in this compound remain to be understood but would further expound on the different spin freezing regimes observed in these spin ice compounds despite the lack of chemical disorder.

References

1. P. Dirac, Quantised singularities in the electromagnetic field. Proc. Roy. Soc. A Math. Phys. Eng. Sci. **133**, 60–72 (1931)
2. P. Milde et al., Unwinding of a skyrmion lattice by magnetic monopoles. Science **340**, 1076 (2013)
3. S.L. Sondhi C. Castelnovo, R. Moessner, Magnetic monopoles in spin ice. Nature **451**, 42 (2008)
4. C. Lacroix, P. Mendels, F. Mila (eds.), *Introduction to Frustrated Magnetism* (Springer, 2011)
5. M.J.P. Gingras, J.S. Gardner, J.E. Greedan, Magnetic pyrochlore oxides. Rev. Modern Phys. **82**, 53 (2010)
6. J.G. Rau, M. Gingras, Spin slush in an extended spin ice model. Nature Communications **7**, 12234 (2016)
7. L. Balents, Spin liquids in frustrated magnets. Nature **464**, 199 (2010)
8. I.A. Ryzhkin, Magnetic relaxation in rare-earth oxide pyrochlores. J. Exp. Theor. Phys. **101**, 481–486 (2005)
9. D.F. McMorrow, T. Zeiske, M.J. Harris, S.T. Bramwell, K.W. Godfrey, Geometrical frustration in the ferromagnetic pyrochlore ho2ti2o7. Phys. Rev. Lett. **13**, 2554 (1997)
10. R.J. Cava, R. Siddharthan, A.P. Ramirez, A. Hayashi, B.S. Shastry, Zero-point entropy in 'spin ice'. Nature **399**, 333 (1999)
11. K. Deguchi R. Higashinaka, H. Fukuzawa, Y. Maeno, Low temperature specific heat of dy2ti2o7 in the kagome ice state. J. Phys. Soc. Jpn. **73**, 2845 (2004)
12. D. Pomaranski et al., Absence of pauling's residual entropy in thermally equilibrated dy2ti2o7. Nature Physics **9**, 353 (2013)
13. S. Rosenkranz et al., Crystal-field interaction in the pyrochlore magnet ho2ti2o7. J. Appl. Phys. **87**, 5914 (2000)
14. R. Siddharthan et al., Ising pyrochlore magnets: Low-temperature properties, "ice rules," and beyond. Phys. Rev. Lett. **83**, 1854 (1999)
15. M. Gingras, B. den Hertog, Dipolar interactions and the origin of spin ice in ising pyrochlore magnets. Phys. Rev. Lett. **84**, 3430 (2000)
16. R. Higashinaka, Y. Maeno, M. Gingras, H. Fukazawa, R. Melko, Magnetic anisotropy of the spin-ice compound dy2ti2o7. Phys. Rev. B **65**, 054410 (2002)
17. T. Fennell et al., Neutron scattering investigation of the spin ice state in dy2ti2o7. Phys. Rev. B **70**, 134408 (2004)
18. J. Snyder et al., Low-temperature spin freezing in the dy2ti2o7 spin ice. Phys. Rev. B **69**, 064414 (2004)
19. Z. Hiroi, K. Matsuhira, T. Sakakibara, T. Tayama, S. Takagi, Observation of a liquid-gas-type transition in the pyrochlore spin ice compound dy2ti2o7 in a magnetic field. Phys. Rev. Lett. **90**, 207205 (2003)

20. T. Fennell et al., Neutron scattering studies of the spin ices ho2ti2o7 and dy2ti2o7 in applied magnetic field. Phys. Rev. B **72**, 224411 (2005)
21. Y. Hinatsu, K. Matsuhira, T. Sakakibara, Novel dynamical magnetic properties in the spin ice compound dy2ti2o7. J. Phys. Condens. Matter **13**, L737 (2001)
22. R.J. Cava, J. Snyder, J.S. Slusky, P. Schiffer, How 'spin ice' freezes. Nature **413**, 48 (2001)
23. E.R. Kassner et al., Supercooled spin liquid state in the frustrated pyrochlore dy2ti2o7. Proc. Natl. Acad. Sci. **112**, 8549 (2015)
24. A.B. Eyvazov et al., Common glass-forming spin liquid state in the pyrochlore magnets dy2ti2o7 and ho2ti2o7. Phys. Rev. B **98**, 214430 (2018)

Chapter 2
Magnetic Monopoles in Spin Ices

The simplest excitations out of the lowest energy 2-in-2-out configuration of the Dy spins in $Dy_2Ti_2O_7$ are not the typically expected spin waves. In fact, individual spin flips breaking the ice rules are the elementary excitations in this compound. These individual flips may be generated by temperature fluctuations, or applied magnetic fields of the order of a few Tesla. Since the pyrochlore lattice is comprised of interconnected tetrahedra, a single spin flip would result in a 3-out-1-in and 3-in-1-out spin configuration on two adjoining tetrahedra (Fig. 2.1 left). The center of a tetrahedron for these positive (negative) defects [1] can be thought of as a source(sink) of magnetic flux and was thus termed as magnetic monopole (anti-monopole) [2, 3] for this solid state system. In this chapter, I discuss the energetics of these monopoles, how they are predicted to interact with each other, as well as some of the past searches for these elusive particles in Dysprosium Titanate. Towards the end of the chapter, a new technique employing a Superconducting QUantum Interference Device is proposed to look for magnetic monopoles.

To understand how these magnetic defects can act like magnetic monopoles, it is important to understand the dumbbell model (Fig. 2.1 right) that inspired this picture. In this model, a Dy spin acting like a magnetic dipole μ can be recast as a dumbbell of opposite signed magnetic charges $\pm q_m = \mu/d$ with separation d. The ratio between diamond lattice constant d and a is : $d = \sqrt{3/2}a = 4.38\text{Å}$. In the limit of this separation tending to zero, the dipolar part of the DSIM is reproduced exactly. The separation between the two charges d is chosen to be the distance between two diamond lattice vortices, described by the centers of tetrahedra in the pyrochlore lattice. Casting each dipole as a dumbbell, the center of each tetrahedron will then contain four dumbbell ends each of charge either $+q_m$ or $-q_m$.

The interaction energy $\mathcal{V}(r_{ij})$ between charges q_i and q_j (residing on sites i,j, separated by distance r_{ij}) can be represented by

© Springer Nature Switzerland AG 2021
R. Dusad, *Magnetic Monopole Noise*, Springer Theses,
https://doi.org/10.1007/978-3-030-58193-0_2

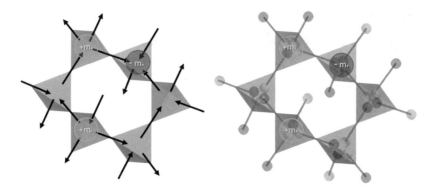

Fig. 2.1 Magnetic monopole charges are created due to spin flips out of the 2-in-2-out lowest energy state of spin ices. The 3-out-1-in / 3-in-1-out spin configurations are sources and sinks of magnetic flux that act as magnetic charges $\pm m_*$ and are free to move about the spin ice lattice left: spin ice model, right: dumbbell model

$$\mathcal{V}(r_{ij}) = \begin{cases} \frac{\mu_0}{4\pi} \frac{q_i q_j}{r_{ij}} & r_{ij} \neq 0 \\ v q_i q_j & r_{ij} = 0 \end{cases} \tag{2.1}$$

The first case represents coulombic interaction between charges $\pm q_m$ and the second term is required to accurately capture the effective exchange interaction between two neighboring dipoles $\pm J_{eff} = \pm (J + 5D)/3$ from the DSIM. It contains a self-energy representing interaction between opposite charges on the same dipole. The value of v can be determined by calculating the interaction energy between two dipoles in the two possible orientations with respect to each other—both dipoles pointing towards the center of a tetrahedron, and one pointing in, the other pointing out and setting this energy equal to J_{eff}.

$$v \left(\frac{\mu}{d}\right)^2 = \frac{J}{3} + \frac{4}{3}\left[1 + \sqrt{\frac{2}{3}} D\right] \tag{2.2}$$

The magnetic charge (m_α) residing at the center of tetrahedron r_α as shown in Fig. 2.1 will then be determined by summing up the four charges $q_{\alpha_1}, q_{\alpha_2}, q_{\alpha_3}, q_{\alpha_4}$. When the spin ice rules are followed by the spins sitting on the vertices of this tetrahedron, i,e., there are two positively charged and two negatively charged dumbbell ends at the center of this tetrahedron, then $m_\alpha = 0$. A monopole is created when the spin-ice rules are violated on a tetrahedron, i.e. when spins are arranged in a 3-out-1-in 3-in-1-out fashion. This magnetic defect in a sea of otherwise spin ice rule following tetrahedra then represents charge of a monopole being $m_\alpha = \pm m_*$. The energy of a magnetic charge configuration containing $m_{\alpha,\beta}$ can then be rewritten in terms of the total charges $\pm m_* = 2\mu/d$ residing on the diamond lattice (defined by tetrahedron centers r_α)

Fig. 2.2 Schematic representation of the spin ice excited state in which two magnetic charges $\pm m_*$ are generated by a spin flip, and propagate through the material. A single flip of an Ising Dy^{3+} spin converts the 2-in/2-out $m_\alpha = 0$ configuration in adjacent tetrahedra, to a situation with adjacent $m_\alpha = \pm m_*$ for 3-out/1-in in one and $m_\alpha = -m_*$ for 3-in/1-out in the next

$$\mathcal{H} = \frac{\mu_0}{4\pi} \sum_{\alpha < \beta} \frac{m_\alpha m_\beta}{r_{\alpha\beta}} + \frac{\nu}{2} \sum_\alpha m_\alpha^2 \qquad (2.3)$$

The first term in the Eq. 2.3 represents a Coulombic interaction between magnetic charges $\pm m_*$. The second, contains ν and enforces the $m_\alpha = 0$ or 2-in-2-out ground state at T=0. In this picture, the cost of creating two neighboring monopoles $\pm m_*$ can be determined from Eq. 2.3 as $\Delta \approx 2(2\nu(\mu/d)^2) - (\frac{\mu_0}{4\pi k_b} \frac{m_\alpha^2}{r_{nn}^2}) = 5.5 K$. An existing 3-out-1-in (3-in-1-out) tetrahedron can be converted to a doubly charged monopoles $\pm 2m_*$ (4-in-1-out and vice versa) with a spin flip of the fourth (out) spin. At temperatures close to the thermal energy barrier for spin flips out of the 2-in-2-out-state $\sim 4.35K$, a plasma or fluid of these $\pm m_*$, with a small population of energetically unfavorable $\pm 2m_*$ charges [4] would exist due to thermal fluctuations.

In general, the magnetic charges in spin ice can move apart via a sequence of spin flips on the tetrahedral network of the pyrochlore lattice (Fig. 2.2). A trail of flipped spins connects the two charges and is colloquially termed as a 'Dirac String', highlighted in yellow in Fig. 2.2. Once a monopole takes a certain path with a Dirac string trailing behind it, another magnetic charge of the same sign cannot sequentially traverse the same path [5]. This is because spin flips required to allow the second monopole of the same sign to traverse a path previously taken are energetically unfavorable. This obstruction is clearly illustrated in Fig. 2.3. This poses constraints on motion of magnetic monopoles in $Dy_2Ti_2O_7$.

2.1 Searches for Monopoles

Since the prediction of the existence of a magnetic monopole fluid was proposed a decade ago, various experiments have looked for these magnetic charges $\pm m_*$ in $Dy_2Ti_2O_7$ [6–12]. A multitude of techniques were employed in these experiments to detect the existence of monopoles—such as study of μSR decay rates, neutron scattering and measurement of susceptibility of magnetic fluid in this material. Two

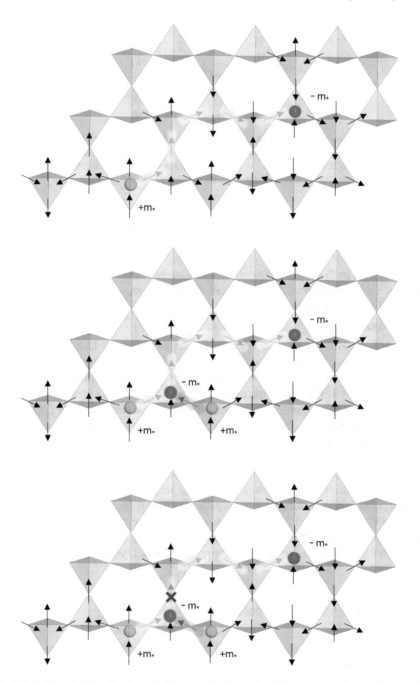

Fig. 2.3 Schematic diagram of constraint on magnetic monopole motion. top: A pair of magnetic monopoles has drifted apart via spin flips and are connected to each other via a Dirac string (highlighted in yellow). middle: A second pair of magnetic monopoles is generated close to the Dirac string. bottom. Magnetic monopole charge of a certain sign is not allowed to sequentially traverse the path taken previously by a magnetic charge of the same sign due to energetically unfavorable spin flips (crossed out spin)

such experiments are discussed here—each involving a study of the response of the magnetic monopole fluid to an applied field.

The first experiment studied neutron scattering spectra from DTO and detected signatures of Dirac strings connecting between monopoles. In zero field, the spectra exhibited pinch points as expected from the coulomb nature of the spin-ice rules (Fig. 2.4a). Next, a strong magnetic field was applied along the [001] direction to magnetize the sample completely and generate a unique ground state. Above 0.6K, when the applied field was reduced to Kasteleyn field $h_K = [k_B T \ln(2)]/(2/\sqrt{3})$, a small fraction of the spins were able to overcome the field and flip with the help of thermal energy. This generated a sparse distribution of Dirac strings executing a random walk. The resultant neutron scattering spectra exhibited cone-like dispersion pattern (Fig. 2.4b), consistent with expectation for diffusion like correlation ($C(x, y, z) \approx \frac{1}{z} \exp(\gamma \frac{x^2+y^2}{z^2})$) among the low density Dirac strings. When the applied field was tilted toward the [011] direction, the random walk of Dirac strings was biased, and the cone of diffusion collapsed onto sheets of scattering (Fig. 2.4c). These observations corroborated with simulations of random walks on a pyrochlore lattice, for both biased and unbiased cases (Fig. 2.4). This experiment showed good agreement between theoretical predictions and experimentally observed signatures of Dirac strings executing random walks in $Dy_2Ti_2O_7$.

Secondly, the prediction of magnetic monopole fluid inspired physicists to measure the flow of this monopole fluid by applying magnetic fields. Two groups approached this problem in distinct ways. The first group measured the AC response of a rod-shaped single crystal of DTO upon the application of a magnetic field [9]. The second experiment performed at Cornell employed boundary free conditions for detection of monopole flow via emf generated from change of magnetization (Fig. 2.5a) in both DTO and HTO [10, 11]. In this experiment, the magnetic 'fluid' was driven by AC and DC fields each and the response of the fluid was measured. While the simple Debye picture [1] of a magnetolyte of monopoles [13] predicted to exist in this system was contradicted by the ac susceptibility experiments (Fig. 2.5b,c), a more nuanced understanding of the dynamics in these spin ice materials was deduced [10, 11].

2.1.1 New Proposal

Since magnetic monopoles in spin ice are deemed to be sources or sinks of magnetic flux, an instrument that can directly detect this flux would be ideal for imaging these charges. A Superconducting QUantum Interference Device (SQUID) is a highly sensitive magnetic flux detector that can be used for this purpose. In fact, such an experiment was carried out in search for the actual Dirac monopole by Cabrera [14]. The principal scheme for such an experiment is described here. A moving fundamental magnetic monopole charge m_0 passing through a superconducting (SC) coil changes the flux through the coil by $\Phi_0 = \pm h/e$, where Φ_0 is the flux quantum.

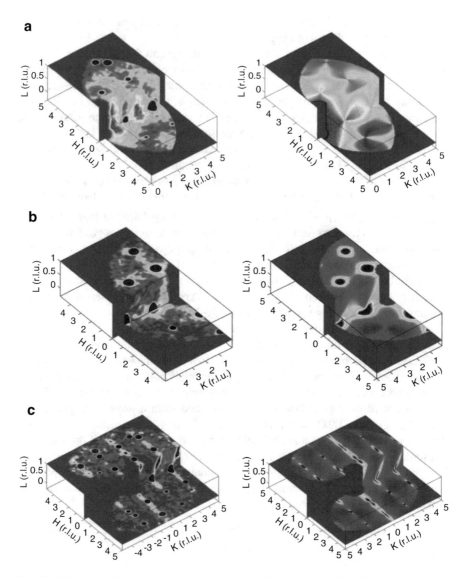

Fig. 2.4 Neutron scattering spectra indicating correlations between Dirac Strings in spin ice compound $Dy_2Ti_2O_7$. (**a**) Observed spectra (left) compared to theoretical prediction of Coulomb gas phase of spin ice in zero field, (**b**) Measured spectra (left) of a sparse density of Dirac strings when field applied along [100] is near Kastelyn transition, stacks up well with predicted spectra from a random walk of Dirac Strings in the pyrochlore lattice (right), (**c**) Neutron scattering spectra measured of a sample in a tilted field towards [110] shows collapse of conical correlations onto sheets and matches theoretical prediction of biased random walks of Dirac strings. From D. J. P. Morris et al. *Dirac strings and magnetic monopoles in the spin ice $Dy_2Ti_2O_7$*. **Science**, 326:411–414 (2009). Reprinted with permission from AAAS

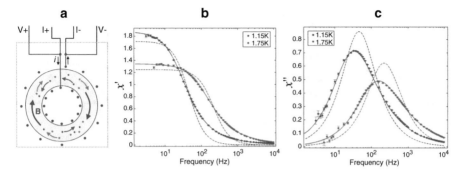

Fig. 2.5 Susceptibility measurements of $Dy_2Ti_2O_7$ in a boundary free geometry of the sample [10]. (**a**) Schematic diagram of the four-probe transport experiment on a boundary free geometry of the $Dy_2Ti_2O_7$ sample. (**b,c**) Real and imaginary parts of ac susceptibility plotted for two temperatures. Dashed lines indicate predictions of simple Debye model for magnetolyte of magnetic monopoles predicted in this material. Solid curves show best fit of more complex magnetic dynamics of Havriliak-Negami form predicted for a supercooled liquid. Figures reproduced with permission from Ref. [10]

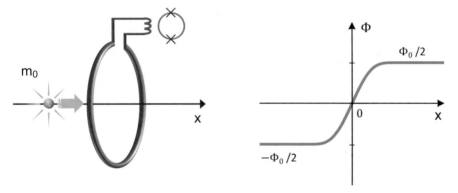

Fig. 2.6 Schematic of fundamental Dirac magnetic monopole with charge m_0 traversing, from $x = -\infty$ to $x = +\infty$, through the input-coil of the SQUID. The magnetic-flux threading the SQUID changes in total by $\Phi_0 = h/e$

Since the ring is SC, the current generated in the ring due to flux change through it does not decay down to zero. This results in a step function jump in the output of the SQUID connected to the SC coil when the magnetic monopole passes through it (Fig. 2.6). One event was detected during the operation of this experiment.

A similar scheme can be applied to detection of emergent monopoles in spin ice materials [2]. If a monopole antimonopole pair with charge $\pm m_*$ is generated at the origin of a SC coil, and the two charges move apart, there's a flux change $\Delta\Phi$ in the SC ring is proportional to their charge $\pm m_*$ given by $\Phi_* = \mu_0 m_*$ (Fig. 2.7). The trail of flipped spins connecting the oppositely charged monopoles acts like the solenoidal Dirac string carrying the flux between the two charges.

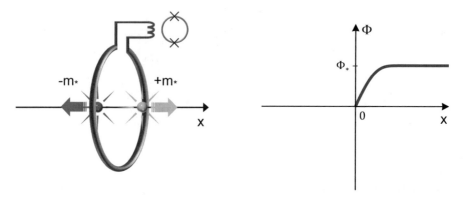

Fig. 2.7 Schematic of two emergent magnetic charges $\pm m_*$ generated in $Dy_2Ti_2O_7$ at $x = 0$ by a thermal fluctuation. As each charge departs in opposite directions to $x = \pm\infty$, the net flux threading the SQUID changes in total by $\Phi_* = \mu_0 m_*$

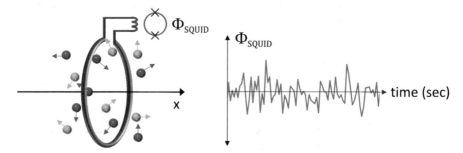

Fig. 2.8 Schematic of a plasma of emergent magnetic charges $\pm m_*$ generated in $Dy_2Ti_2O_7$ by thermal fluctuations. Instead of a step function signal, we would expect to observe noise from the plethora of magnetic charges threading the SQUID pickup coil

In the temperature range of 1K-4K, close to the thermal energy barrier of 4.35K for spin flips in DTO, the density of monopoles is high in a macroscopic sample of mm size and the magnetic flux signal from such a sample might be expected to appear as stochastic noise $\Phi(t)$ as shown schematically in Fig. 2.8. The magnitude of noise is expected to depend on temperature, as monopoles are generated through thermal spin flips. To search for such magnetic flux noise from monopoles we built a highly sensitive spin noise spectrometer that uses a SQUID.

References

1. I.A. Ryzhkin, Magnetic relaxation in rare-earth oxide pyrochlores. J. Exp. Theor. Phys. **101**, 481–486 (2005)
2. S.L. Sondhi C. Castelnovo, R. Moessner, Magnetic monopoles in spin ice. Nature **451**, 42 (2008)

3. R. Moessner, C. Castelnovo, S. Sondhi, Spin ice, fractionalization, and topological order. Annu. Rev. Condens. Matter Phys. **3**, 35 (2012)
4. V. Kaiser et al., Emergent electrochemistry in spin ice: Debye-hückel theory and beyond. Phys. Rev. B **98**, 144413 (2018)
5. P. Holdsworth, L. Jaubert, Magnetic monopole dynamics in spin ice. J. Phys. Condens. Matter **23**, 164222 (2011)
6. D.J.P. Morris et al., Dirac strings and magnetic monopoles in the spin ice dy2ti2o7. Science **326**, 411–414 (2009)
7. S.T. Bramwell et al., Measurement of the charge and current of magnetic monopoles in spin ice. Nature **461**, 956 (2009)
8. D. Prabhakaran, G. Aeppli, S.T. Bramwell, L. Bovo, J. Bloxsom, Brownian motion and quantum dynamics of magnetic monopoles in spin ice. Nature Communication **4**, 1535 (2013)
9. Yaraskavitch et al., Spin dynamics in the frozen state of the dipolar spin ice material dy2ti2o7. Phys. Rev. B **85**, 020410 (2012)
10. E.R. Kassner et al., Supercooled spin liquid state in the frustrated pyrochlore dy2ti2o7. Proc. Natl. Acad. Sci. **112**, 8549 (2015)
11. A.B. Eyvazov et al., Common glass-forming spin liquid state in the pyrochlore magnets dy2ti2o7 and ho2ti2o7. Phys. Rev. B **98**, 214430 (2018)
12. C. Paulsen et al., Experimental signature of the attractive coulomb force between positive and negative magnetic monopoles in spin ice. Nature Physics **12**, 661 (2016)
13. R. Moessner, C. Castelnovo, S. Sondhi, Debye-hückel theory for spin ice at low temperature. Phys. Rev. B **84**, 144435 (2011)
14. B. Cabrera, First results from a superconductive detector for moving magnetic monopoles. Phys. Rev. Lett. **48**, 1378 (1982)

Chapter 3
Experiment

The objective of the experiment presented in this thesis is to perform direct detection of magnetic monopoles in $Dy_2Ti_2O_7$. This could be achieved by optimizing the sensitivity of detector to flux noise from a plethora of monopoles anticipated to be present in mm-sized samples of $Dy_2Ti_2O_7$ in the temperature range of 1K-4K. To be able to detect noise coming from monopole motion alone, it was important to eliminate all other sources of noise from entering the detector. To that effect, a custom cryostat was built to house the spectrometer. This chapter details the design of a 1K cryostat and the apparatus that houses the Dysprosium Titanate sample that comprise the Spin Noise Spectrometer. Initial observations of flux noise observed in our experiment and the premise for Monte Carlo simulations of spin fluctuations in Dysprosium Titanate are also presented. Finally, a new paradigm for understanding magnetic monopoles: generation recombination noise is introduced.

3.1 1K Cryostat

The operation of a 1K cryostat is based on the concept of evaporative cooling which has been known to mankind since time immemorial. The boiling point of liq. Helium is $\sim 4.2K$ and pumping on this liquid in a closed system reduces its boiling point. Cooling of the experiment is done in different stages. First, the experimental apparatus consisting of a vacuum chamber is brought down to 4K by placing the vacuum can (VC) in a bath of liquid helium. Then a capillary continuously brings in a small volume of liquid helium from the bath to a copper pot inside the VC. Once the pot fills up with liquid helium, it is pumped upon via a stainless steel tube with diameter 3/4 in. This pumping lowers the temperature of the pot and connected experiment down to $\approx 1.2K$. The volume of the pot has been optimized to be 25 cm^3, A schematic diagram of the 1K cryostat in operation is shown in Fig. 3.1. The

© Springer Nature Switzerland AG 2021
R. Dusad, *Magnetic Monopole Noise*, Springer Theses,
https://doi.org/10.1007/978-3-030-58193-0_3

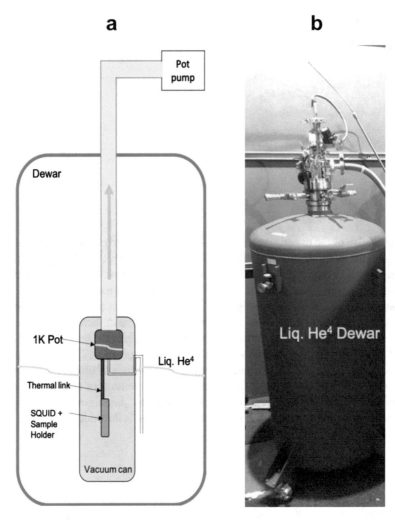

Fig. 3.1 (a) Schematic diagram of 1K cryostat, (b) Photograph of 1K cryostat in operation

'insert' for the 1K cryostat with the experiment mounted at the base of the 1K pot is shown in Fig. 3.2.

The main experimental setup consisting of a DC SQUID Model 550 chip from Quantum Design and sample holding geometry is connected to the base of the pot. To ensure optimal thermalization of the experiment to a desired set temperature, the SQUID chip assembly is held with a thick chunk of brass. The chunk of brass is then connected to the base of the pot via three thin C-shaped brass connectors to ensure optimal thermal conductance. The temperature of the experiment is controlled using PID tools onboard Lakeshore 340 with the help of a thermometer and heater placed close to the experiment as shown in Fig. 3.3.The thermometer used in the experiment is a Lakeshore CERNOX placed close to the sample by mounting it on the brass

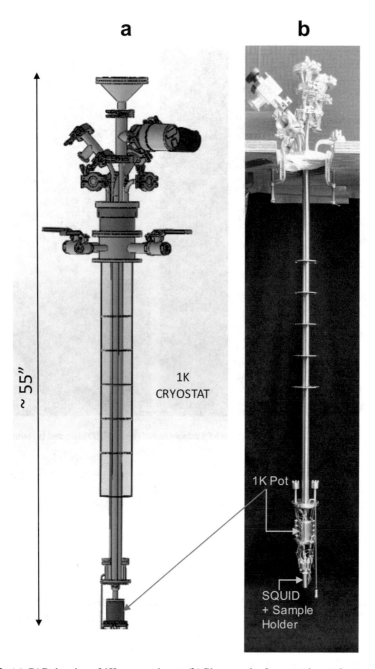

Fig. 3.2 (**a**) CAD drawing of 1K cryostat insert, (**b**) Photograph of cryostat insert shown

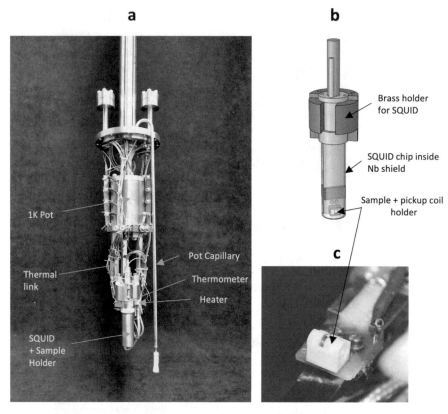

Fig. 3.3 (**a**) View of 1K Pot that is thermally connected to (**b**). SQUID chip and (**c**) sample holder

assembly holding the SQUID + sample assembly. The heater consists of a 100 Ohm metal film resistor, again placed next to the SQUID+ sample assembly.

3.2 Spectrometer

The spin noise spectrometer (SNS) setup (Fig.3.4) consists of cylindrical sample-holder (SH), a 'saddle' that holds the SH and mounts it atop the SQUID chip, and the SQUID circuitry. Commercially available QD 550 SQUID from Quantum Design is used to construct the SNS. The SH has a concentric hole of diameter 1.4 mm and length 5.7 mm to hold samples. The design of the SH is modular as samples of varying geometries that can fit inside the concentric hole of the SH are easily replaceable with the help of tweezers. Generally, rod-shaped samples of $Dy_2Ti_2O_7$ are studied in this experiment. A superconducting pickup coil consists of 6 turns of thin NbTi wire of diameter ~ 0.09 mm is closely wound around the outside of the

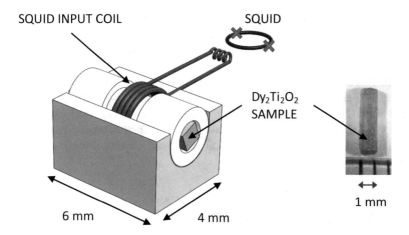

SQUID INPUT COIL   SQUID

$Dy_2Ti_2O_2$
SAMPLE

6 mm   4 mm

1 mm

Fig. 3.4 Schematic representation of Spin Noise Spectrometer: MACOR cylindrical shell with concentric hole for rod-shaped samples, pickup coil is placed on the outside of the shell, a saddle shaped MACOR part mounts the sample holder onto SQUID chip

SH. This pickup coil is connected to the input coil of the SQUID via Nb crimp pads. The inductance of the pickup coil ($L \approx 0.25 \mu H$) is optimized to match that of the input coil of the SQUID to maximize coupling between the two coils.

3.2.1 External Noise Insulation of Spectrometer

The SQUID chip and SH assembly is situated inside a strong flux shield made of a Nb tube with an aspect ratio $R \sim 4$. This electromagnetic shielding and SQUID circuitry is optimized by the manufacturer such that the noise floor of the bare SQUID chip is $\sim 3\mu\phi_0/\sqrt{Hz}$ for a bandwidth of 2.5kHz. To insulate the detector from external noise, no electrical cables are allowed to enter the shielded SQUID chip setup. The cryostat is placed in an acoustically shielded sound room which is separate from the room that houses the pot pump to minimize the effect of mechanical vibrations on the experiment. The measures mentioned here insure that flux-noise floor of this spectrometer sits at $\delta\Phi < 4\mu\phi_0/\sqrt{Hz}$ in the entire temperature range of study (shown in Appendix A.1).

3.2.2 Sample Preparation

The $Dy_2Ti_2O_7$ samples investigated in this experiment were grown in an optical floating zone furnace in Prof. Graeme M. Luke's Group at McMaster University. X-ray diffraction on the resultant crystal was sharp and showed no signs of twinning or the presence of multiple grains. Performance of Rietveld refinement on the diffraction data yields a unit cell lattice constant of 10.129 Å; this implies a maximum possible level of "spin stuffing" (substitution of Dy_{3+} ions on Ti_{4+} sites) of $\sim 2.9\%$ and a most likely spin-stuffing fraction $< 1\%$.

Disk shaped $Dy_2Ti_2O_7$ crystals obtained from Prof. Luke' group were then cut into rod-shaped samples with the help of a dicing saw encrusted with diamond bits on the blade. From previous transport experiments [1] on $Dy_2Ti_2O_7$ crystals it was learned that orientation of these crystals does not affect magnetic response of the material. Therefore, no particular crystal axis direction was chosen when the rod-shaped samples were cut.

3.2.3 Recording Sample Spin Fluctuations

A typical operation cycle of the SNS aboard the 1K cryostat consists of cooling down experiment to 1.2K and then using PID controls to vary temperature from 1.2K to 4K with temperature stability at each point of 2.5 mK. Once the temperature is stable at a desired set-point, spin fluctuations inside DTO sample are detected by the pickup coil connected to the SQUID input coil. The DC SQUID used in this SNS operates in a flux-locked loop supported by the electronics on the SQUID chip and preamplifier. The detected flux is output from the SQUID electronics box as an analog voltage $V_{SQUID}(t)$ calibrated to the flux detected by a transfer function (Appendix C.2).

To perform spectral analysis of the signal, $V_{SQUID}(t)$ is fed into a SR780 Dynamic Signal Analyzer with input noise of $\sim 300nV/\sqrt{Hz}$, much below that of the SQUID noise floor. The SA then calculates the autocorrelation function $C(\tau)$ of $V_{SQUID}(t)$ and then takes the Fourier transform of $C(\tau)$ to generate a power spectrum (averaged) $S_v(\omega, T)$ that is sent to a PC to be to be recorded. A schematic diagram of spin noise detection scheme is shown in Fig. 3.5. Unprocessed data at each temperature consists of 5 datasets of bandwidth (BW) 2.5kHz, each of which is an outcome of averaging of 1000 acquisitions where acquisition time = (1/resolution BW), resulting in the measured SQUID output as voltage-flux noise spectral density detected at the SQUID. This power spectrum $S_v(\omega, T)$ is then converted to $S_\Phi(\omega, T)$ using calibration described in Appendix C.2.

3.3 Initial Observations

The first temperature controlled noise measurements of $Dy_2Ti_2O_7$, showed that flux noise from rod shaped samples of Dysprosium Titanate displayed a strong dependence on temperature and frequency. The noise spectrum at 4K had a distinct shape—a plateau followed by an eventual decay. The decay could be characterized by a cutoff frequency, that shifted to a lower value as the sample was cooled.

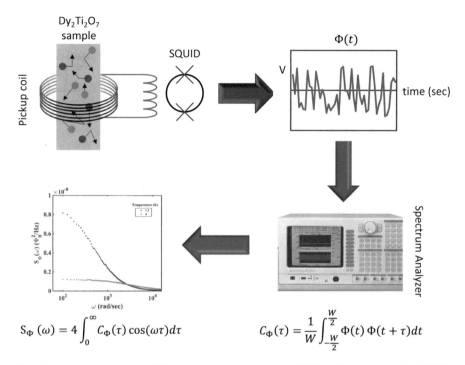

$$S_\Phi(\omega) = 4\int_0^\infty C_\Phi(\tau)\cos(\omega\tau)d\tau \qquad C_\Phi(\tau) = \frac{1}{W}\int_{-\frac{W}{2}}^{\frac{W}{2}}\Phi(t)\,\Phi(t+\tau)dt$$

Fig. 3.5 Schematic diagram of flux noise detection using a SQUID. The output of the SQUID is voltage V(t) which is calibrated to flux detected $\Phi(t)$ by the SQUID. This output is fed into a spectrum analyzer that calculates the autocorrelation function $C_\Phi(\tau)$ and then the noise spectral density of the flux $S_\Phi(\omega)$

Detection of magnetic monopoles by measuring spin noise of $Dy_2Ti_2O_7$ was proposed [2] shortly after our first experiments were conducted. Kirschner et al. calculated noise of stray field at a distance 10 nm from a $Dy_2Ti_2O_7$ sample with Monte Carlo (MC) simulations of the DSIM. Figure 1 of Ref. [2] compares spin noise from three different spin ice models at two temperatures: 1K and 4K. We noted that DSIM predicted noise for $Dy_2Ti_2O_7$ had similar temperature and frequency dependence to our experimental observation (Fig. 3.6), and established a collaboration with Prof. S. Blundell to pursue the MC theory relevant to our experiment.

3.4 Monte Carlo Simulations

The Monte Carlo study simulated thermally generated magnetic configurations of $Dy_2Ti_2O_7$ from the Dipolar Spin Ice Model as presented in Eq. 1.1 at a given temperature and then modeled the spin flip dynamics (Fig. 3.7). Taking into account

Fig. 3.6 Top: Raw data
$S_\Phi(\omega, T)$ from $Dy_2Ti_2O_7$
measured by our spin noise
spectrometer at two
temperatures. Bottom: MC
calculated noise from stray
field of a $Dy_2Ti_2O_7$ sample
for different spin ice
hamiltonians. Reprinted
figure with permission from
F. Kirschner et al., Phys. Rev.
B., 97, 140402 (2018)
Copyright (2018) by the
American Physical Society
[2]

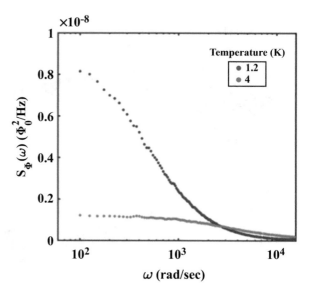

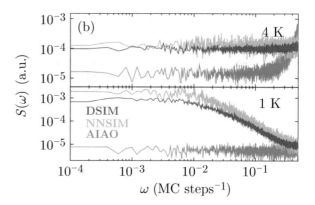

the fact that our measurement was concerned with the bulk of the material, the
simulation scheme calculated magnetic fields from the bulk of a $Dy_2Ti_2O_7$ sample
instead of stray field a distance away from the sample. In general, these simulations
were carried out on a sample containing 4x4x4 unit cells, each of which contains
16 Dy^{3+} ions. This is referred to as the MC sample henceforth. Standard MC
procedures were used [3], consisting of 10^6 cooling steps followed by an interval
of 5000 MC-time-steps, at fixed T. During this interval W, the time dependence
of net z-component of magnetic moment $\mu_Z(t)$ of the whole MC sample is then
simulated. This procedure is then repeated 600 times. The range of temperatures of
these simulations was between 4.0K and 1 K. Because this is a simulation of bulk
magnetization dynamics, periodic boundary conditions were used in all directions.

Fig. 3.7 Visual
representation of Monte Carlo
simulation of Dy spin lattice
containing 4x4x4 unit cells of
$Dy_2Ti_2O_7$ (courtesy
Franziska Kirschner). The
3-in-1-out /3-out-1-in spin
configurations are labeled
with red and blue spheres at
the center of those tetrahedra

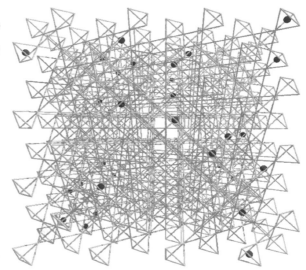

Precisely, for a given conformation, the z-component of magnetic moment of
the MC sample $\mu_Z(t)$ was found by summing z-components over the individual
magnetic moments of the 1024 Dy spins, where $\mu \approx 10\mu_B$. The simulated time
dependence of this value at a given temperature T is $\mu_Z(t, T)$, and is evaluated
sequentially during the time window W. Its autocorrelation function is

$$C_{\mu_Z}(\tau, T) = \frac{1}{W} \int_{-W/2}^{W/2} \mu_Z(t, T)\mu_Z(t + \tau, T)dt \, [\mu_B^2] \qquad (3.1)$$

The predicted spectral density of magnetization noise in the MC sample is then
calculated using the Wiener-Khinchin theorem

$$S_{\mu_Z}(\omega, T) = 4 \int_0^\infty C_{\mu_Z}(\tau, T)cos(\omega\tau)d\tau \, [\mu_B^2 MCstep] \qquad (3.2)$$

We extract the frequency range $10^{-4}(MC - steps)^{-1}$ to $10^{-1}(MC - steps)^{-1}$ (the
Nyquist frequency is $0.5(MC - steps)^{-1}$). Equation 3.2 was then averaged over the
600 independent simulation runs to the get better precision for $S_{\mu_Z}(\omega, T)$.

To bring the calculated noise density into more universal units, we use
$S_{M_Z}(\omega, T) = (\mu_B^2)S_{\mu_Z}(\omega, T))/V^2 \, [A^2m^{-2}MCstep]$ where $V = 6.6 \times 10^{-26}m^3$
is the volume of the MC sample. In order to compare the magnitude of noise
density from experiment and theory, an estimate of the same from a sample and
pickup coil with dimensions comparable to our experiment were required to be
made. Based on an understanding of sample geometric effects, it was estimated
that $N = 2.9 \cdot 10^{16} \pm 20\%$ MC samples are present in the experimental volume.
Therefore $S_{M_Z}(\omega, T)$ was divided by N following statistics of stochastic processes.

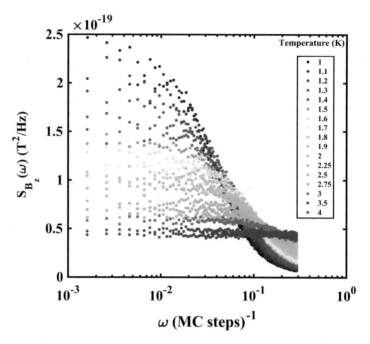

Fig. 3.8 MC estimate of $S_{B_z}(\omega, T)$ from a $Dy_2Ti_2O_7$ sample of equivalent volume to our experimentally studied sample

The spectral density of fluctuations of z-component of magnetic field B[Tesla] within the sample is then

$$S_{B_z}(\omega, T) = \mu_o^2 S_{M_z}(\omega, T)[T^2 MCstep] \qquad (3.3)$$

Figure 3.8 then shows an approximate estimate of $S_{B_z}(\omega, T)$ for our specific sample geometry calculated by MC simulations using the DSIM (Eq. 1.1).

3.5 Paradigm Shift from Resistor to Semiconductor

With a prediction for magnetic field noise spectral density at hand (Fig. 3.8), spin noise spectroscopy of rod-shaped $Dy_2Ti_2O_7$ samples was conducted in the temperature range $1.2K \leq T \leq 4K$, and the calibrated $S_\Phi(\omega, T)$ is reported in Fig. 3.9. The general trends of frequency and temperature seen in the experimentally measured flux noise spectral density from $Dy_2Ti_2O_7$ and MC calculated field noise were very similar. However, the rise of noise plateau height with fall in temperature was quite puzzling (Figs. 3.9 and 3.10).

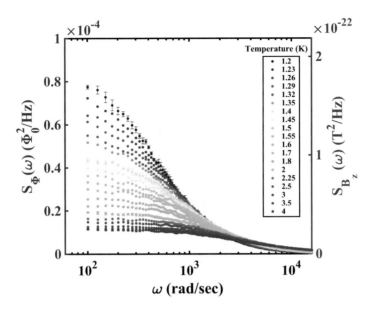

Fig. 3.9 Calibrated $S_\Phi(\omega, T)$ from our $Dy_2Ti_2O_7$ sample in the entire T range of $1.2K \leq T \leq 4K$. Right axis shows equivalent $S_{B_z}(\omega, T)$ for our sample generated by using the cross-section of our sample

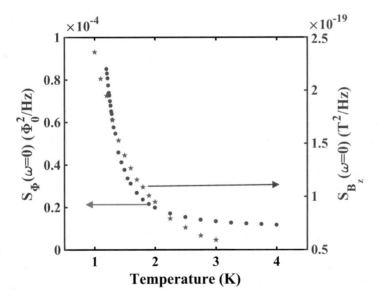

Fig. 3.10 Plateau height of both measured $S_\Phi(\omega, T)$ (left), and MC estimated $S_{B_z}(\omega, T)$ (right) rise with a fall in temperature

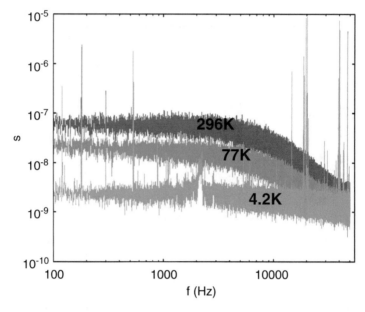

Fig. 3.11 Measured thermal noise of an electrical resistor at three different temperatures. Figure reproduced from Ref. [4]

Naively one might anticipate magnetic monopoles to exhibit thermal noise akin to electronic Johnson noise (Fig. 3.11)

$$S_V(\omega, T) = \frac{4k_B T R}{1 + \omega^2 \tau^2} \tag{3.4}$$

Here $S_V(\omega, T)$ is the open-circuit voltage fluctuations in a resistor R as a function of temperature T and frequency ω.

However our experimental observation was strikingly different from Johnson noise. The low temperature dataset in our experiments sat at a higher plateau height than the high temperature dataset (Fig. 3.10), whereas the opposite is true for thermal noise from a resistor. This suggested that perhaps a different mechanism was needed to explain the microscopic mechanism behind magnetization noise we were observing.

Noise from generation and recombination (GR) of electrons and holes in semiconductors exhibits unanticipated temperature dependence. The plateau height of this electronic GR noise fell with rise in temperature [5]. The temperature and frequency dependence of such GR noise [5] looked similar to experimentally measured flux noise from $Dy_2Ti_2O_7$ (Fig. 3.12).

Thermal noise in a resistor is generated by random motion of electrons due to their kinetic energy which is directly proportional to temperature. Generation-recombination noise occurs due to thermally stimulated generation of electron-hole

Fig. 3.12 Generation
Recombination noise
measured in a Semiconductor.
©(1969) IEEE. Reprinted,
with permission, from IEEE
Transactions on Electron
Devices, 16, 170–177, 1969
[5]

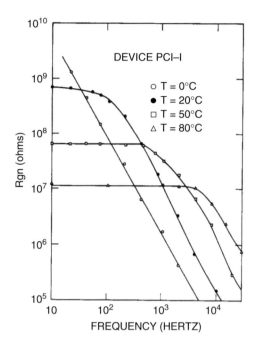

pairs in a semiconductor and subsequent recombination of electrons and holes. The
rate of generation and recombination of electron and holes are also temperature
dependent. The key difference between the structure of electric noise originating
from two mechanisms is that the height of plateau of thermal noise of a resistor
decreases with temperature, whereas, somewhat counter-intuitively, the height
of plateau of GR noise increases with decreasing temperature. This induced a
paradigm shift in thinking about Johnson noise of magnetic monopoles to noise
from generation and recombination of magnetic monopoles. In the next chapter, we
explore the mapping of statistics of electron-hole generation recombination [6, 7],
onto magnetic monopoles [8].

References

1. E.R. Kassner et al., Supercooled spin liquid state in the frustrated pyrochlore dy2ti2o7. Proc. Natl. Acad. Sci. **112**, 8549 (2015)
2. A. Yacoby, N. Yao, F.K.K. Kirschner, F. Flicker, S.J. Blundell, Proposal for the detection of magnetic monopoles in spin ice via nanoscale magnetometry. Phys. Rev. B **97**, 140402 (2018)
3. A. Konczakowska, B.M. Wilamowski, *Fundamentals of Industrial Electronics, Chapter 11* (Taylor and Francis, 2011)
4. L. Motta, B. Wilson, Johnson noise: Phys 504 lab - Yale University (2007). http://yalelab. wikidot.com/results

5. C-T. Sah, L.D. Yau, Theory and experiments of low-frequency generation-recombination noise in mos transistors. IEEE Trans. Electron. Dev. **16**, 170 (1969)
6. R.E. Burgess, The statistics of charge carrier fluctuations in semiconductors. Proc. Phys. Soc. B **69**, 1020 (1956)
7. C.M. Wilson, D.E. Prober, Quasiparticle number fluctuations in superconductors. Phys. Rev. B **69**, 094524 (2004)
8. M. Ryzhkin, A. Klyuev, A. Yakimov, Statistics of fluctuations of magnetic monopole concentration in spin ice. Fluctuation Noise Lett. **16**, 1750035 (2017)

Chapter 4
Plasma of Magnetic Monopoles

In the past decade of trying to understand how potential magnetic monopoles would behave in this material, the focus was on the 'free'/independent motion of this magnetic charge. However, it was important to keep in mind and realize that these charges come in pairs of opposite signs. A system similar to this has been in existence for over half a century and is the basis of modern electronics. To understand the temperature and frequency dependence of noise from a magnetic monopole system consisting of equal and opposite charges, it could be instructive to look to its electronic cousin—the semiconductor with electrons and holes. In an intrinsic semiconductor, electric charges $\pm q$ that are subject to Coulomb interactions may also undergo spontaneous generation and recombination processes (Fig. 4.1a) that are well understood [1–3]. Here, thermal generation and recombination (GR) of $\pm q$ pairs generates a spectral density of voltage noise $S_V(\omega, T) = V^2 S_N(\omega, T)/N_0^2$, where $S_N(\omega, T)$ is the spectral density of GR fluctuations in the number of $\pm q$ pairs. In this chapter the formalism for magnetic monopole antimonopole generation and recombination is developed. The predictions coming out of this formalism for the analytic form of magnetic monopole noise are layed out. Finally, Monte Carlo simulations for Dysprosium Titanate in the temperature range of 1.2K-4K are compared with the aforementioned predictions.

There is a direct analogy between GR of electric hole pairs $\pm q$ and magnetic $\pm m_*$ charges (Fig. 4.1). Magnetic monopoles are spontaneously generated via thermal spin flips and can recombine at a later time, again with the help of thermal spin flips. The similarity between generation-recombination in electronic semiconductors and magnetic monopole housing spin ices was utilized to predict the magnetic spectral noise density for magnetic monopoles [4]. Here we describe the statistics of generation-recombination noise arising from thermally activated magnetic monopole plasma of equal and opposite charges.

© Springer Nature Switzerland AG 2021
R. Dusad, *Magnetic Monopole Noise*, Springer Theses,
https://doi.org/10.1007/978-3-030-58193-0_4

a

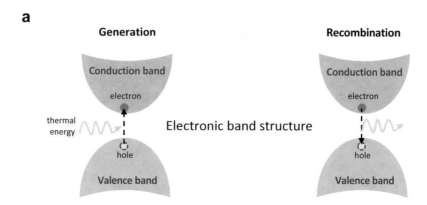

b

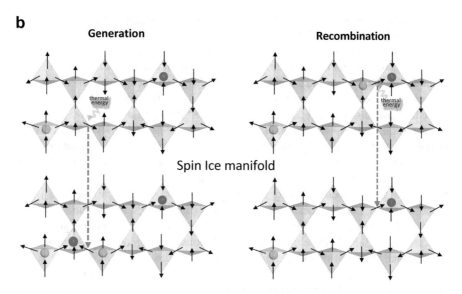

Fig. 4.1 (**a**) Schematic representation of electronic semiconductor generation of electron-hole pair stimulated by thermal energy and a recombination process leading to a release of thermal energy. (**b**) Schematic diagram of both generation and recombination of magnetic monopoles being stimulated by thermal energy in a spin ice manifold

4.1   Magnetic Monopole Generation and Recombination Noise

We define the number of monopole-antimonopole pairs N at a temperature T as $N(T)$. At low temperatures, we expect most of the spins in spin ice to follow the ice rules. Some of the spins flip out of the 2-in-2-out state due to thermal stimulus and monopoles (3-in-1-out/3-out-1-in) are generated at a rate $g(N, T)$.

These monopoles can move about the lattice freely and can recombine with other monopoles of the opposite charge at a rate of $r(N, T)$. The master equation describing the probability $P(N, T)$ of finding N monopoles at a certain temperature can be written as follows

$$\frac{dP(N, T)}{dt} = r(N + 1, T)P(N + 1, T) + g(N - 1, T)P(N - 1, T)$$

$$- P(N, T)[g(N, T) + r(N, T)] \qquad (4.1)$$

Here $g(N, T)$ and $r(N, T)$ represent the generation and recombination rates of the monopoles. One pair of monopoles is added or removed by a generation or recombination event respectively. At thermal equilibrium there exists a most probable value of number of monopoles at that temperature $N(T) = N_0(T)$ that stays constant. We note that in steady state, the rate of generation of these monopoles will equal the rate of recombination of these monopoles. The exact dependence of g and r on N and T depends on the microscopics of generation and recombination process pertaining to the specific system under investigation.

$$g(N_0, T) = r(N_0, T) \qquad (4.2)$$

Thermal fluctuations can make the system move out of equilibrium momentarily by changing the number of monopoles by δN so that $\delta N = N - N_0$. From the Eq. 4.1, the Langevin Equation for these magnetic charge number fluctuations can been derived

$$\frac{d\langle \delta N \rangle}{dt} = -\frac{\langle \delta N \rangle}{\tau(T)} + \sqrt{A}\zeta(t) \qquad (4.3)$$

Here, τ represents the time constant for N to approach its equilibrium value after a fluctuation has occurred and $A \propto g(N_0)$, and $\zeta(t)$ represents the thermally generated stimulus uncorrelated in time that has a normalized spectrum such that $S_\zeta(f) = 1$ [4]. The GR rate τ can be defined as

$$\frac{1}{\tau(T)} = \frac{d(r - g)}{dN}\bigg|_{N_0} = r'(N_0, T) - g'(N_0, T) \qquad (4.4)$$

Taking the Fourier transform of Eq. 4.3 and taking an ensemble average yields the predicted spectral density of $\pm m_*$ pair fluctuations as

$$S_N(\omega, T) = \frac{\sigma_N^2(T)\tau(T)}{1 + \omega^2\tau^2(T)} \qquad (4.5)$$

where σ_N^2 is the variance in the number of $\pm m_*$ pairs.

4.2 Variance of Monopole Pair Number Fluctuations

We can deduce that generation recombination $\tau(T)$ behaves in a manner that matches our expectations from previous experiments. Decoding what $\sigma_N(T)$ for magnetic monopoles will shed further light on what Eq. 4.5 can reveal about magnetic monopole noise. From Eq. 4.1, expanding $\ln P(N, T)$ about its maximum value $\ln P(N_0, T)$ in a quadratic fashion [1, 2] yields

$$\left[\frac{\partial^2}{\partial N^2} \ln P(N, T)\right]_{N=N_0} = \frac{g'(N_0, T)}{g(N_0, T)} - \frac{r'(N_0, T)}{r(N_0, T)} \tag{4.6}$$

$$\ln P(N, T) = \ln P(N_0, T) - \frac{1}{2}(N - N_0)^2 \left[\frac{r'(N_0, T)}{r(N_0, T)} - \frac{g'(N_0, T)}{g(N_0, T)}\right] \tag{4.7}$$

Thus the expected Gaussian probability distribution of N about its most probable value N_0 is

$$P(N, T) = P(N_0, T) \exp\left[-(N - N_0)^2 / 2\overline{(N - N_0)^2}\right] \tag{4.8}$$

The variance of monopole number $\sigma_N^2 = \overline{(N - N_0)^2}$ is then determined from Eqs. 4.6 and 4.8 [1, 2] as

$$\sigma_N^2(T) = \left[\frac{r'N_0, T)}{r(N_0, T)} - \frac{g'(N_0, T)}{g(N_0, T)}\right]^{-1}$$

$$= \frac{g(N_0, T)}{r'(N_0, T) - g'(N_0, T)} \tag{4.9}$$

$$= g(N_0, T) \cdot \tau(T)$$

4.2.1 Variance for Monopole Number Fluctuations in Dysprosium Titanate

For emergent magnetic monopoles in $Dy_2Ti_2O_7$, the equilibrium generation rate at temperature T within 1.2K and 4K will approximately be $g(N_0, T) \propto \exp(-\Delta/T)$ [5] where Δ is the thermal energy barrier for spin flips required to generate monopoles. It is established from previous experiments that at these temperatures, that the time constants are given approximately by $\tau(T) \sim \exp(\Delta/T)$ [6]. This implies that the variance of magnetic monopole number $\sigma_N^2 \propto \exp(\Delta/T) \cdot \exp(-\Delta/T)$ should be approximately constant in this temperature range.

4.3 Predictions

With the tools we have developed in this chapter and the calculations of $S_{B_z}(\omega, T)$ from MC, it was now possible to predict the structure of magnetic monopole noise coming from a $Dy_2Ti_2O_7$ sample that is studied in our experiment.

1. **Analytic form**

    Monopole number fluctuations arising out of slight differences in generation and recombination rates of these particles will manifest as fluctuations in flux through a superconducting coil. We predict that $S_N(\omega, T)$ (as shown in Figs. 4.2 and 4.3) will embody itself in $S_\Phi(\omega, T)$ measured by our detector.

$$S_\Phi(\omega, T) \propto \frac{\tau(T)}{1 + \omega^2\tau^2(T)} \tag{4.10}$$

Fig. 4.2 Predicted spectral density of fluctuations in monopole number $S_N(\omega, T)$ from Eq. (4.5), for several monopole GR time constants τ, in a range that one might expect to achieve by cooling $Dy_2Ti_2O_7$ from 4K to $\sim$1K. The GR plateau in $S_N(\omega, T)$ is clear as $\omega \to 0$ as is the ω^{-2} falloff expected of free monopole motion for frequencies $\omega\tau > 1$

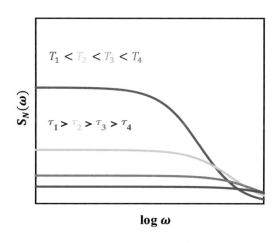

Fig. 4.3 Normalized spectral density $S_N(\omega, T)/S_N(0, T)$ shows variation of microscopic time constant τ with temperature

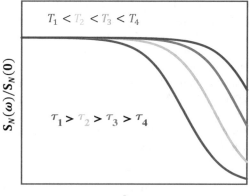

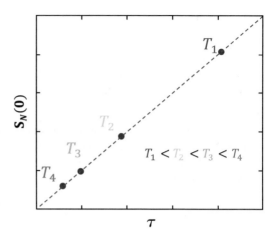

Fig. 4.4 For magnetic monopole GR with $\sigma_N(T)$ constant as a function of temperature, $S_N(\omega = 0, T) \propto \tau(T)$

2. **GR time constant**

The microscopic time constant $\tau^{-1}(T) = \left.\frac{d(r-g)}{dN}\right|_{N_0}$ describing GR processes in Dy$_2$Ti$_2$O$_7$ can be obtained by analyzing $S_\Phi(\omega, T)$ as can be seen in Fig. 3.9.

3. **Plateau height of Noise vs time constant**

As derived for magnetic monopoles, $\sigma_N(T)$ is approximately a constant for $1.2K \leq T \leq 4K$. When this property of variance of magnetic monopole number is input into Eq. 4.5, it is found that for $\omega = 0$, $S_N(0, T) \propto \tau(T)$ as is shown in Fig. 4.4.

4.4 Comparison with MC Calculations

MC simulations of magnetic noise $S_{B_z}(\omega, T)$ from a Dy$_2$Ti$_2$O$_7$ sample by employing DSIM (Eq. 1.1) demonstrates that spin noise spectroscopy can be used to directly detect magnetic monopole generation and recombination predicted in DTO. These simulations also describe how the microscopic time constants in the material, albeit in MC steps, can vary as a function of temperature. The form of $S_{B_z}(\omega, T)$ revealed by MC studies (Fig. 4.5) is equivalent in its key characteristics to Eq. 4.5 (Fig. 4.2). Here the relationship $S(\omega) \propto \omega^{-2}$ that holds true at high frequencies for a single GR time constant in GR theory, is replaced with a more nuanced behavior $S(\omega) \propto \omega^{-b}$ with $b(T) < 2$. To explore this enhancement of spin noise from Dy$_2$Ti$_2$O$_7$, MC simulations of $B_z(t)$ are carried out for a total of three different models of $\pm m_*$ plasma, and characteristics of $S_{B_z}(\omega, T)$ from these are compared to the experiment. These models are described below

1. **Dipolar Spin Ice**

The dipolar spin ice model (DSIM) leads to a lowest energy state of Dy spins pointing in 2-in-2-out state on each tetrahedron (Fig. 1.2). As previously discussed, the violation of this rule by a spin flip causes generation of a monopole

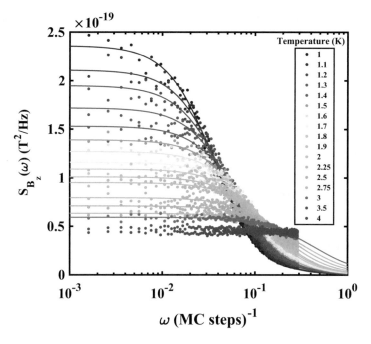

Fig. 4.5 Predicted spectral density of magnetic field fluctuations within the $Dy_2Ti_2O_7$ sample $S_{B_z}(\omega, T)$ from MC simulations using the Hamiltonian in Eq. 1.1 in a range $1K \leq T \leq 4K$. The solid lines represent the functional form $S_{B_z}(\omega, T) \propto \tau(T)/(1 + (\omega\tau(T))^{b(T)})$ that was used to fit the MC predictions

anti-monopole pair with charge $\pm m_*$ or doubly charged pair with charge $\pm 2m_*$ [7]. The energy of two nearest-neighbor monopoles is 3.06K, and the energy to create one monopole is $\Delta = 4.35K$. Since the spins sit on tetrahedral corners, magnetic monopole motion is guided by spin flips in a topologically constrained fashion. These monopoles experience a strong coulombic force between the $\pm m_*$ charges.

2. **Nearest Neighbor Spin Ice**

   The nearest-neighbor spin ice (NNSI) Hamiltonian is considered by setting D=0 in Eq. 1.1. It suppresses the effects of long-range coulombic interactions. J is chosen such that the system still has a 2-in-2-out ground state, while having the same density of excitations as DSI at a given temperature. This system still has monopole-like excitations, but greatly reduced force between the monopoles.

3. **Free plasma**

   Free plasma refers to a system of magnetic charge pairs $\pm m_*$ moving freely in the absence of Coulomb interactions or topological constraints due to the Dirac strings in $Dy_2Ti_2O_7$. The model is specified in Eq. 2.3 with $\pm m_*$ charges located on the sites of a diamond lattice.

In the next chapter, the above listed predictions are tested with respect to experimentally measured $S_\Phi(\omega, T)$ shown in Fig. 3.9.

References

1. K.M. van Vliet, J.R. Fassett, *Fluctuation Phenomena in Solids, edited by R. E. Burgess* (Academic Press, New York, 1965)
2. R.E. Burgess, The statistics of charge carrier fluctuations in semiconductors. Proc. Phys. Soc. B **69**, 1020 (1956)
3. L. Reggiani, V. Mitin, L. Varani, *Noise and Fluctuations Control in Electronic Devices, Chapter 2* (American Scientific Publishers, 2002)
4. M. Ryzhkin, A. Klyuev, A. Yakimov,   Statistics of fluctuations of magnetic monopole concentration in spin ice. Fluctuation Noise Lett. **16**, 1750035 (2017)
5. R. Moessner, C. Castelnovo, S.L. Sondhi, Debye-hückel theory for spin ice at low temperature. Phys. Rev. B **84**, 14435 (2011)
6. K. Matsuhira et al., Spin dynamics at very low temperature in spin ice dy2ti2o7. J. Phys. Soc. Jpn. **80**, 123711 (2011)
7. V. Kaiser et al., Emergent electrochemistry in spin ice: Debye-hückel theory and beyond. Phys. Rev. B **98**, 144413 (2018)

Chapter 5
Analysis

The experiment we conduct is based on the premise that monopoles traversing the pickup coil of a SQUID will thread flux Φ_* through it, proportional to their charge $\pm m_*$. Instead of step function jumps in the SQUID signal (Fig. 2.7), a magnetic noise $\Phi(t)$ (Fig. 2.8) is expected to be detected by the highly sensitive flux noise spectrometer developed for this experiment. This noise originates from a thermally stimulated dense plasma of monopoles that get generated and may recombine in the temperature range $1.2K \leq T \leq 4K$. In this chapter, the experimental results of Spin Noise Spectroscopy of Dysprosium Titanate are presented and analyzed. These results are then compared to the predictions made by the analytic formulation of generation recombination noise as well as Monte Carlo simulations of spin noise coming from a $Dy_2Ti_2O_7$ sample.

Our measurements of this magnetic noise $S_\Phi(\omega, T)$ have revealed a strong temperature and frequency dependence. We discover that the noise has a distinct shape—a plateau at low frequencies ($\omega\tau \ll 1$) and an eventual decay at higher frequencies ($\omega\tau \gg 1$). The inflection point for this noise ($\omega\tau \approx 1$) can reveal details about the microscopic time constant in the monopole motion through the material. We find that the temperature dependence of the detected $S_\Phi(\omega, T)$ is counter-intuitive.

5.1 Analytic Structure of Noise

Flux noise from a sample of $Dy_2Ti_2O_7$, which is theoretically presumed to host a plasma of magnetic monopoles, is measured via spin noise spectroscopy. To check if noise originating from generation and eventual recombination of magnetic monopoles as outlined in the previous chapter best explains the structure of our

© Springer Nature Switzerland AG 2021
R. Dusad, *Magnetic Monopole Noise*, Springer Theses,
https://doi.org/10.1007/978-3-030-58193-0_5

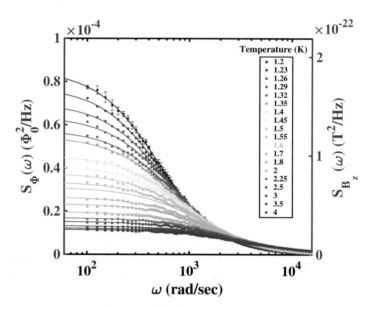

Fig. 5.1 Measured spectral density of flux-noise $S_\Phi(\omega, T)$ from a $Dy_2Ti_2O_7$ sample in the range $1.2K \leq T \leq 4K$. The left-hand axis is the magnetic-flux noise spectral density $S_\Phi(\omega, T)$; the right-hand axis is an estimate of the equivalent magnetic-field noise spectral density $S_{B_z}(\omega, T)$ averaged over the $Dy_2Ti_2O_7$ samples. The best fit to the function $\tau(T)/(1 + (\omega\tau(T))^{b(T)})$ shown as a fine solid curve. Overall we find $S_\Phi(\omega, T)$ of $Dy_2Ti_2O_7$ to be constant for frequencies 1Hz $< f(T) = \frac{1}{2\pi\tau(T)}$, above which it falls off as ω^b

measurements, each data set $S_\Phi(\omega, T)$ was fit to the Eq. 4.10 with the best fit shown as a solid curve in Fig. 5.1. The free parameters for the fits were : GR time constant $\tau(T)$, $S_\Phi(0, T)$, and b. The fit qualities are excellent at all temperatures with $R^2 > 0.99$. The residuals for these fits and further details of the data analysis are shown in Appendix A.3. We find that the magnetic-flux noise spectral density of $Dy_2Ti_2O_7$ is constant for frequencies from near 1Hz up to an angular frequency $w(T)$ $1/\tau(T)$, above which it falls off as ω^{-b} where b spans a range between 1.2 and 1.5. We find that the overall characteristics of GR noise are retained in the magnetic noise spectrum observed. The behavior of each of these parameters is discussed in the following text.

5.2   Extraction of Time Constant

When variation of $S_\Phi(\omega, T)$ is examined in the frequency domain (Fig. 5.1), it is found that the inflection point for the noise plateau changes with temperature. A qualitative understanding of this phenomenon can be gained by looking at the normalized noise spectra $S_\Phi(\omega, T)/S_\Phi(0, T)$. This function is plotted in Fig. 5.2 and

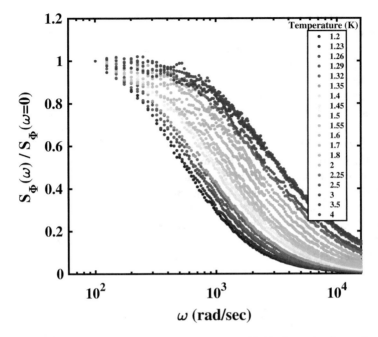

Fig. 5.2 Normalized spectral density of flux noise $S_\Phi(\omega, T)/S_\Phi(0, T)$ coming from a $Dy_2Ti_2O_7$ sample, revealing the divergence of the time constant $\tau(T)$ toward longer times at lower temperatures

it is apparent that $\tau(T)$ evolves rapidly toward longer times at lower temperatures. The shape of normalized spectra is identical to that predicted by analytical GR noise shown in Fig. 4.3.

To quantitatively study how the GR time constant varies with temperature we now focus on $\tau(T)$ extracted by fitting $S_\Phi(\omega, T)$ to Eq. 4.10. From this procedure, we find that $\tau(T)$ diverges at low temperatures (Fig. 5.3), indicating freezing. The rate of this divergence is not Arrhenius ($\tau(T) = A \exp(\Delta/T)$) as is expected for a thermally activated process, but described by the Volger-Tammann-Fulcher (VTF) equation where D is not Dipolar coupling, but a dimensionless constant

$$\tau(T) = \tau_0 \exp\left(\frac{DT_0}{T - T_0}\right) \tag{5.1}$$

In previous susceptibility experiments, the relaxation time constant of $Dy_2Ti_2O_7$ was studied by measuring the response of the magnetization of the material to an applied field. It has been established empirically that the susceptibility-derived microscopic time-constants $\tau_M(T)$ involved in magnetic dynamics of $Dy_2Ti_2O_7$ diverge with decreasing T [1–3] and are heterogeneous [4, 5]. There have been many ac susceptibility measurements of $Dy_2Ti_2O_7$ in different sample shapes ranging from polycrystalline samples [2] to toroidal single crystals [4, 5]. Yaraskavitch et al. established that the time constants $\tau_M(T)$ measured from SQUID-based susceptibility measurements of rod shaped samples were in good qualitative agreement with $\tau_M(T)$ reported in both in Snyder et al. and Matsuhira et al. Eyvazov et al. verified quantitative agreement between $\tau_M(T)$ from their susceptibility measurements and $\tau_M(T)$ from Yaraskavitch et al.

From Fig. 5.3 we can see that $\tau(T)$ obtained from our flux-noise experiments clearly follows a quantitatively equivalent trajectory to the ac susceptibility $\tau_M(T)$ of Refs. [4] and [5] which also exhibit a VTF form. The VTF parameters for the $\tau(T)$ obtained from flux noise measurements and ac susceptibility experiments of Eyvazov et al. are shown in Table 5.1. Thus our measurements of $\tau(T)$ corresponds well with the $\tau_M(T)$ derived from numerous susceptibility studies that are widely cited for this material [1–5]. This correspondence between $\tau_M(T)$ and $\tau(T)$ remains to be understood at the level of quantitative microscopic theory. Theoretical calculations involving quantum tunneling describing spin flips in $Dy_2Ti_2O_7$ at these temperatures may shed further light on this issue [6].

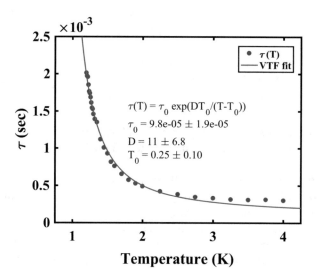

Fig. 5.3 Plot of time constant from fits to measured $S_\Phi(\omega, T)$ data as shown in Fig. 5.1. The flux-noise derived time constant behaves in a super Arrhenius fashion

Table 5.1 Table comparing the $\tau(T)$ VTF parameters for SNS and AC susceptibility experiments [5]

Measurement	τ_0 (sec)	D	T_0 (K)
Flux noise	9.8×10^{-5}	11	0.25
AC susceptibility	1.4×10^{-4}	14	0.26

5.3 Linear Relationship Between $S(0, T)$ and $\tau(T)$

One of the most intriguing observations arising from measurements of $S_\Phi(\omega, T)$ was that that plateau height of the noise increased as T decreased (Fig. 3.10). GR noise of magnetic monopoles directly predicts the growth of $S_\Phi(\omega = 0, T)$ with falling temperature as a natural consequence of $S_\Phi(\omega = 0, T) \propto \tau(T)$, and $\tau(T)$ diverging at low temperatures in DTO. In Fig. 5.4 measured $S_\Phi(0, T)$ is plotted against measured $\tau(T)$ from fits in Fig. 5.1 where T is the implicit variable for the temperature range of our experiment. Thus we find that $S_\Phi(0, T) \propto \tau(T)$ throughout the full T range.

The relationship between the $S_\Phi(\omega = 0, T)$ and $\tau(T)$ is a result of $\sigma_N(T)$ in GR noise of monopole number fluctuations in DTO being approximately a constant as a function of T in the range of 1.2K-4K. Kluyev et al. make a similar assumption while working out an expression for $S_N(\omega)$, however a physical reasoning for this is not provided in Ref. [7].

5.3.1 Variance of Monopole Flux Noise

The quantity $\sigma_\Phi^2(T)$ is measured at each temperature by integrating measured monopole flux noise $S_\Phi(\omega, T)$ with respect to frequency for the entire bandwidth.

$$\sigma_\Phi^2 = \int_0^\infty S_\Phi(\omega, T)d\omega \qquad (5.2)$$

Fig. 5.4 $S_\Phi(0, T)$ plotted versus $\tau(T)$ as measured from fitting data in Fig. 5.1. Observation that $S_\Phi(0, T) \propto \tau(T)$ for $Dy_2Ti_2O_7$ throughout the full temperature range is a key expectation for $\pm m_*$ GR magnetic-flux noise

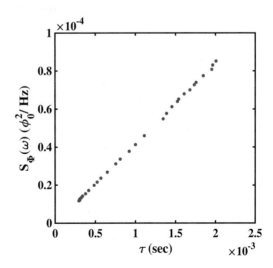

Fig. 5.5 Plot of measured variance of flux σ_Φ^2 shows that it is approximately constant as a function of temperature, in the entire temperature range of our experiment

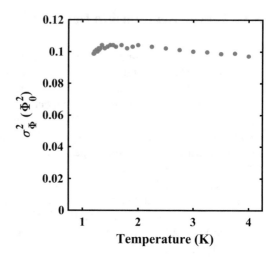

From Fig. 5.5, it is noted that σ_Φ^2 is weakly dependent on temperature. This validates monopole GR theory prediction for $\sigma_N^2(T)$. Figure 1.8 suggests that different activation processes describe the dynamics of spin ices, and that $\tau(T)$ across a wide temperature range has different behaviors. A deeper understanding of why σ_Φ^2 is approximately constant in the range of 1.2K–4K would be revealed by extending the temperature range of the experiment and knowledge of the exact mechanism of spin flips [6] determining the generation and recombination rate of monopoles.

5.4   Comparison with MC Calculations

Finally we examine our experimental measurements in the context of MC calculations of magnetic field noise coming from $Dy_2Ti_2O_7$. The MC study of magnetic field fluctuations arising in a sample of $Dy_2Ti_2O_7$ by simulating the spin flips according to DSIM (Eq. 1.1) is an important and profound prediction for magnetic field noise we should expect to see in our experiment.

To compare the MC calculations and experiment on the same footing, it would be useful to have x-axis of Fig. 3.8 in actual time units. This is done by converting from MC-step to seconds (described in Appendix C.2), so that angular frequency for MC ω(rad/sec)=$2\pi/t$(sec) and $S_{B_z}(\omega, T)$ [T^2 s].

From Fig. 5.6 we see that both experiment $S_\Phi(\omega, T)$, and MC $S_{B_z}(\omega, T)$ from a $Dy_2Ti_2O_7$ sample retain the overall characteristic predictions of GR noise (Fig. 4.2) as is apparent from respective fits to the functional form $S_B(\omega, T) \propto \tau/(1+(\omega\tau)^b)$.

It is now possible to test the prediction of $S(\omega = 0, T) \propto \tau(T)$ for MC magnetic field noise from magnetic monopole GR theory. When $S_{B_z}(0, T)$ is plotted versus $\tau(T)$ (Fig. 5.7 bottom), they are approximately proportional, once an offset to all values of $S_{B_z}(0, T)$ due to numerical Nyquist (sampling) noise is considered.

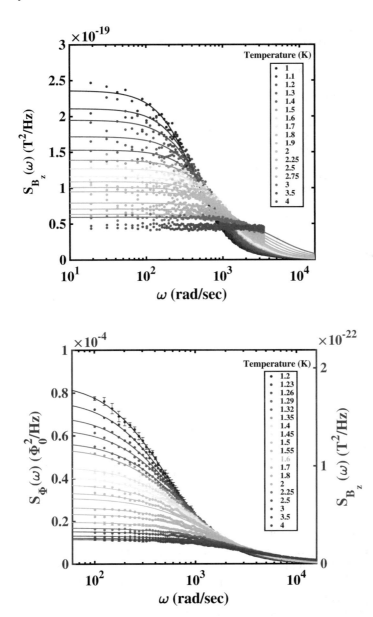

Fig. 5.6 Top: Predicted spectral density of magnetic field fluctuations within the $Dy_2Ti_2O_7$ sample $S_{B_z}(\omega, T)$ from MC simulations using the Hamiltonian in Eq. 1.1 in a range $1K \leq T \leq 4K$. Bottom: Measured spectral density of flux-noise $S_\Phi(\omega, T)$ from $Dy_2Ti_2O_7$ samples n the range $1.2K \leq T \leq 4K$. The left-hand axis is the magnetic-flux noise spectral density $S_\Phi(\omega, T)$; the right-hand axis is an estimate of the equivalent magnetic-field noise spectral density $S_{B_z}(\omega, T)$ averaged over the $Dy_2Ti_2O_7$ sample

Fig. 5.7 Top: Predicted relationship from $Dy_2Ti_2O_7$ MC simulations of $B_Z(t)$, of $S_{B_z}(0, T)$ versus $\tau(T)$ for the GR fluctuations of a $\pm m_*$ magnetic charge plasma. Note that all $S_{B_z}(0, T)$ are offset by a constant along the y-axis due to artifacts of Nyquist (sampling) noise at the high frequency end of the MC calculations. Bottom: $S_\Phi(0, T)$ plotted versus $\tau(T)$ as measured from fitting data in Fig. 5.1. Observation that $S_\Phi(0, T) \propto \tau(T)$ for $Dy_2Ti_2O_7$ throughout the full temperature range is a key expectation for $\pm m_*$ GR magnetic-flux noise

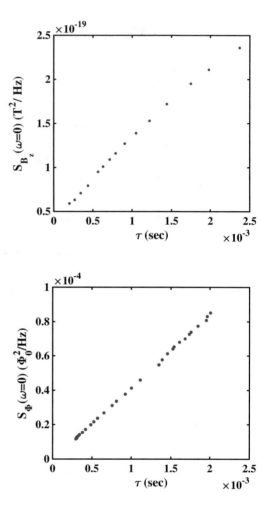

We note that the size of MC calculated magnetic field noise spectra differs from the experimentally measured $S_{B_z}(\omega, T)$ by three orders of magnitude. Since the scaling from MC simulation to experimental mm size sample involves a volume ratio of 10^{16}, it is hardly surprising that the flux noise magnitudes detected are different than expected by this amount. While finite size scaling of MC simulations for DSI Hamiltonians has been done in the past [8], estimating the magnitude of the flux noise to be expected for a mm-scale sample of $Dy_2Ti_2O_7$ is a challenge for future MC simulations.

From the GR fits, it is found that the the power law of frequency for decay of GR noise, i.e. $b(T)$ in $S_B(\omega, T) \propto \tau/(1 + (\omega\tau)^b)$ is less than 2 for both experiment and MC. The modified $b(T)$ suggests new physics that cannot be described by a simple theory of free magnetic monopole plasma that has been derived so far [7]. We discuss implications of this revelation in the next chapter.

References

1. J. Snyder et al., Low-temperature spin freezing in the dy2ti2o7 spin ice. Phys. Rev. B **69**, 064414 (2004)
2. K. Matsuhira et al., Spin dynamics at very low temperature in spin ice dy2ti2o7. J. Phys. Soc. Jpn. **80**, 123711 (2011)
3. Yaraskavitch et al., Spin dynamics in the frozen state of the dipolar spin ice material dy2ti2o7. Phys. Rev. B **85**, 020410 (2012)
4. E.R. Kassner et al., Supercooled spin liquid state in the frustrated pyrochlore dy2ti2o7. Proc. Natl. Acad. Sci. **112**, 8549 (2015)
5. A.B. Eyvazov et al., Common glass-forming spin liquid state in the pyrochlore magnets dy2ti2o7 and ho2ti2o7. Phys. Rev. B **98**, 214430 (2018)
6. R. Moessner, J. Quintanilla, B. Tomasello, C. Castelnovo, Brownian motion and quantum dynamics of magnetic monopoles in spin ice (2018). arXiv:1810.11469
7. M. Ryzhkin, A. Klyuev, A. Yakimov, Statistics of fluctuations of magnetic monopole concentration in spin ice. Fluctuation Noise Lett. **16**, 1750035 (2017)
8. R.G. Melko, M.J.P. Gingras, Monte carlo studies of the dipolar spin ice model. J. Phys. Condens. Matter **16**, R1277 (2004)

Chapter 6
Correlations in Magnetic Monopole Motion

To understand the correlations in $\pm m_*$ GR noise, we compare our experimental knowledge to $S_{B_z}(\omega)$ MC predictions made by three different spin interaction hamiltonians for spin ices as explained in Sect. 5.4. By varying certain parameters like dipolar coupling D or constraints on $\pm m_*$ motion, we learn about how the magnetic monopole noise is more complex than that of a free plasma of $\pm m_*$ charges.

6.1 Power-Law Exponent b

The MC predictions of exponent b for the power-law falloff of magnetic-flux noise from the three theories DSI (blue), NNSI (green) and free monopoles (red) can be determined by fitting each simulated $S_{B_z}(\omega, T)$ to $\tau(T)/(1 + (\omega\tau(T))^{b(T)})$. The results are shown in Fig. 6.1. Experimentally measured $b(T)$ from fitting to $S_{B_z}(\omega, T)$ in Fig. 5.1 are shown as black dots.

We find that amongst the three $\pm m_*$ dynamics models studied, DSIM predictions for the exponent b, are most consistent with the experimental measurements of b. This indicates that both topological constraints (lacking in the free plasma model) and strong dipolar interactions between spins (suppressed in NNSI) play important roles in magnetic monopole dynamics in $Dy_2Ti_2O_7$. The slight difference between DSIM $b(T)$ and experimental $b(T)$ suggests that the correlated monopole dynamics is more complex than can be anticipated by available MC simulations. A fully accurate Hamiltonian for the spin dynamics of $Dy_2Ti_2O_7$ or $Ho_2Ti_2O_7$ is known to more complex than the spin-ice Hamiltonian of Eq. 1.1, with significant effects from correlations [1]. However, simulation of the dynamics in such complex Hamiltonians, that may require more near neighbor exchange parameters, exceeds the numerical capacity of MC simulations today.

© Springer Nature Switzerland AG 2021
R. Dusad, *Magnetic Monopole Noise*, Springer Theses,
https://doi.org/10.1007/978-3-030-58193-0_6

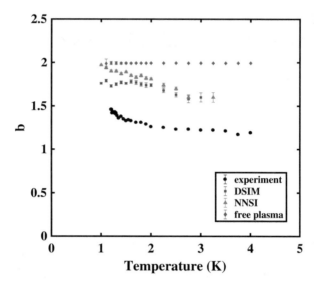

Fig. 6.1 Monte-Carlo prediction of exponent b in $S_{B_z}(\omega, T) \propto \tau(T)/(1 + (\omega\tau(T))^{b(T)})$ for the three magnetic charge dynamics theories. These are the DSI model (blue); the NNSI model (green); the free plasma model (red). Measured exponent b from fitting $S_\Phi(\omega, T) \propto \tau(T)/(1 + (\omega\tau(T))^{b(T)})$ for all data in Fig. 5.1 is shown in black. The time constants $\tau(T)$ of the MC $S_{B_z}(\omega, T)$ and of the $S_\Phi(\omega, T)$ data are not free parameters here

6.2 Autocorrelation Function

Monte-Carlo simulations for $Dy_2Ti_2O_7$ can directly predict the autocorrelation function $C_{B_z}(t, T)$ of magnetic field fluctuations $B_z(t)$ [2] as described in Sect. 3.4. Figure 6.2 shows $\log[C_{B_Z}(t, T)/C_{B_Z}(0, T)]$ predictions for three distinct magnetic charge dynamics theories at T=1.2K. The first MC model (blue) describes $\pm m_*$ magnetic charge plasma of dipolar spin ice (DSI) which has Coulomb-like inter-particle interactions, and in which the existence of a Dirac-string (yellow Fig. 2.2) produces very strong constraints by preventing another monopole of the same charge from following the same route (Fig. 2.3) [2, 3]. The second MC model (green) is the nearest neighbor spin ice model (NNSI) in which Coulomb-like inter-particle interactions are absent but Dirac-string constraints present. Our final model (red) is a neutral plasma of $\pm m_*$ magnetic charges that is topologically unconstrained. For comparison, the measured autocorrelation function $\log[C_{B_Z}(t, T)/C_{B_Z}(0, T)]$ of magnetic-field fluctuations $B_Z(t)$ is plotted in black and with a best-fit curve overlaid.

Clearly, the DSI model, including Coulomb-like interactions and Dirac-string topological constraints, is far more consistent with measured correlations in this system. Moreover, the NNSI model which lacks the Coulomb-like interactions, is inconsistent with the experiment, and short time correlations appear to be completely absent. Except for evolution of the time constant $\tau(T)$, these correlation

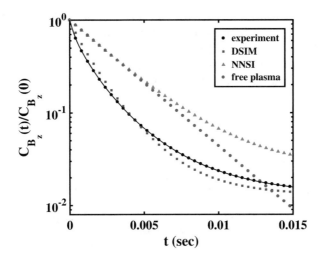

Fig. 6.2 For T=1.2K, the Monte-Carlo simulation prediction of the autocorrelation function $\log[C_{B_z}(t)/C_{B_z}(0)]$ in the field fluctuations $B_z(t)$ for three models of the spin dynamics of $Dy_2Ti_2O_7$. The DSI model contains Coulomb-like interactions and constraints on repeated passage of a same-sign monopoles along the same trajectory due to Dirac strings (green); the nearest neighbor spin ice model (NNSI) has had the Coulomb interactions suppressed (blue); the free monopole plasma (red) is in-keeping with free monopole GR theory. The measured autocorrelation function $\log[C_{B_z}(t)/C_{B_z}(0)]$ of magnetic-field fluctuations $B_z(t)$ of $Dy_2Ti_2O_7$ is plotted in black and overlaid with a fit function; the experimental error bars are smaller than the data points. Clearly the autocorrelation function of the DSI model corresponds best to the measured $C_{B_z}(t)$. We note that the distinction between the single slope (red) for the free monopole plasma, and the more complex predicted $C_{B_z}(t)$ for the other cases, represents microscopically a distinction between a simple process involving a single time constant vs. a more complex one, potentially involving a spread of relaxation time constants. Most importantly, the measured $C_{B_z}(t)$ (black) shows that magnetization dynamics is obviously strongly correlated in time

phenomena are virtually unchanged within our temperature range. We note that the distinction between the single slope (red) for the free monopole plasma, and the more complex predicted $C_{B_Z}(t)$ for the other cases, represents microscopically a distinction between a simple process involving a single time constant vs. a more complex one. Most importantly, the measured $C_{B_Z}(t)$ (black) shows that magnetization dynamics is obviously strongly correlated in time.

We note that the curved shape of $C_{B_Z}(t)$ for experiment and DSIM (Fig. 6.2 black and blue dataset) indicates heterogeneous timescales. Kluyev et al. predict that a distribution of microscopic spin relaxation timescales should result in a combination of noise spectra, $S_i(\omega) \propto \tau_i/(1 + (\omega\tau)^2)$, for each time scale τ_i in the distribution. The result would be a total noise spectrum $S(\omega) \propto \tau/(1 + (\omega\tau)^b)$ where τ is a central value and b is controlled by the relative weights of the different τ_i.

Comparison between simulated and measured autocorrelation functions $C_{B_z}(t, T)$ and falloff exponents b, for magnetic-flux noise in $Dy_2Ti_2O_7$ reveals that the DSI model is most consistent with the observed phenomenology. To achieve precise agreement may require adjustment of the J, D terms [1] in Eq. 1.1, or better control over finite size scaling effects [4]. But overall, the data in Figs. 6.1 and 6.2 imply that the power-law signatures of strong correlations observed in both $\log[C_{B_z}(t)/C_{B_z}(0)]$ and $S_{B_z}(\omega, T)$ are occurring due to a combination of the existence of a Dirac-string trailing each monopole (Fig. 2.2) and the Coulombic interactions.

References

1. T. Fennell et al., Neutron scattering investigation of the spin ice state in dy2ti2o7. Phys. Rev. B **70**, 134408 (2004)
2. A. Yacoby, N. Yao, F.K.K. Kirschner, F. Flicker, S.J. Blundell, Proposal for the detection of magnetic monopoles in spin ice via nanoscale magnetometry. Phys. Rev. B **97**, 140402 (2018)
3. P. Holdsworth, L. Jaubert, Magnetic monopole dynamics in spin ice. J. Phys. Condens. Matter **23**, 164222 (2011)
4. R.G. Melko, M.J.P. Gingras, Monte carlo studies of the dipolar spin ice model. J. Phys. Condens. Matter **16**, R1277 (2004)

Chapter 7
Fluctuation Dissipation Theorem

In our experiment, we have measured spin noise spectrum of a Dysprosium Titanate
sample at equilibrium, from 1.2K to 4K using a DC-SQUID. From the Fluctuation
Dissipation (FD) theorem, we know that statistical fluctuations in a physical variable
of a system are related to the linear response to a small force applied to the system.
In this chapter our experimental results of flux noise coming from a sample of
$Dy_2Ti_2O_7$ are discussed in the context of previous boundary-free AC susceptibility
measurements of Dysprosium Titanate.

If we take the system as magnetic monopole fluid in $Dy_2Ti_2O_7$, the power
spectrum of the fluctuations of magnetic field in this fluid is related to $\chi''(\omega, T)$,
the imaginary part of susceptibility of the magnetic fluid to small external forces

$$S_{B_z} \propto \frac{\chi''(\omega, T) \cdot k_B T}{\omega} \tag{7.1}$$

With the SNS, we have measured the flux noise spectral density $S_\Phi(\omega, T) \approx$
$\langle \Phi^2(\omega, T) \rangle$ generated due to equilibrium thermal fluctuations in Dysprosium
Titanate. This flux noise $S_\Phi(\omega, T)$ is proportional to magnetic field noise $S_{B_z}(\omega, T)$
picked up by our spectrometer.

$$S_\Phi(\omega, T) = \langle \Phi^2(\omega, T) \rangle = \sigma^2 S_{B_z} \tag{7.2}$$

where σ is the area of the SQUID input coil. Using our measurements of $S_\Phi(\omega, T)$
and equivalent $S_{B_z}(\omega, T)$ (Eq. 7.2) for a $Dy_2Ti_2O_7$ sample, $\chi''(\omega, T)$ was determined (Fig. 7.1) in arbitrary units (Eq. 7.1).

This measured $\chi''(\omega, T)$, extracted from flux noise measurements was then fit
to the Havriliak Negami form of $\chi''(\omega, T)$ that Kassner et al. [1] ac susceptibility
experiment determined as

$$\chi(\omega, T) = \chi_\infty + \frac{\chi_0}{(1 + (i\omega\tau)^\alpha)^\gamma} \tag{7.3}$$

© Springer Nature Switzerland AG 2021
R. Dusad, *Magnetic Monopole Noise*, Springer Theses,
https://doi.org/10.1007/978-3-030-58193-0_7

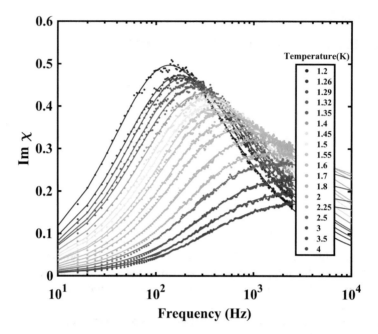

Fig. 7.1 Imaginary part of susceptibility of a fluid of monopoles in $Dy_2Ti_2O_7$ in arbitrary units, extracted from our $S_\Phi(\omega, T)$ measurements, vs frequency is plotted here for temperature range of 1.2K to 4K

Free parameters for the fitting procedure were χ_0, τ, α and γ. The fits are of good quality, with $R^2 > 0.99$ for all temperatures studied (Fig. 7.2). We have demonstrated some elements of the FD theorem for $Dy_2Ti_2O_7$ samples by using our spin noise studies to predict imaginary part of susceptibility for this material. The functional form of $\chi''_\Phi(\omega, T)$ is equivalent to that of $\chi''_M(\omega, T)$ determined by Kassner et al. [1].

7.1   Sample Geometry Effects

There is a slight difference between τ_Φ and τ_M and might be present due to sample geometries in the two experiments being quite different. It will require quantitative tests to establish a relationship between geometry of sample and the microscopic time constant τ_Φ (extracted from fit shown in Fig. 7.1). As noted in some previous transport measurements [2], edges of a rod-shaped magnetic sample create demagnetizing stray fields that are picked up by a detector.

Shape effects could occur in such spin noise measurements. This is because even though we are measuring a cuboidal sample with a coil around the middle, spins at the ends of the sample still contribute partially. Our experiments measure flux

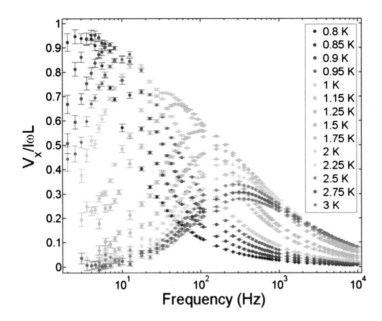

Fig. 7.2 Imaginary part of susceptibility for a $Dy_2Ti_2O_7$ sample vs frequency from Kassner et al. is plotted here for temperature range of 0.8K to 3K. Figure reproduced with permission from from Ref. [1]

through the pickup coil due to the dipole fields from spins in the sample, and the noise is coming fundamentally from spin flips (aka monopole hops). The flux at the pickup coil due to a single spin in the sample depends on where that spin is and which way it is pointing. Some spins are invisible (e.g. spins pointing to a direction in the xy plane) and some have a greater effect, e.g. spins close to, but not at the edge, in or near the plane of the coil, pointing along z. If there is a fluctuation of the magnetic moment of the whole sample, resulting in a net moment, then that would produce a net demagnetization field which all the spins would experience, potentially affecting their dynamics, and that demagnetization field would be shape-dependent.

References

1. E.R. Kassner et al., Supercooled spin liquid state in the frustrated pyrochlore dy2ti2o7. Proc. Natl. Acad. Sci. **112**, 8549 (2015)
2. Yaraskavitch et al., Spin dynamics in the frozen state of the dipolar spin ice material dy2ti2o7. Phys. Rev. B **85**, 020410 (2012)

Chapter 8
Conclusions

Noise is an entity that most scientists attempt to sideline in their measurements. There may however be some benefits to examining the characteristics of noise intrinsic to a black box that one is studying. A famous example of the usefulness of such a study is the discovery of the Cosmic Microwave Background that stemmed from an unprecedented measurement [1]. In this thesis, the intrinsic noise coming from a single crystal of $Dy_2Ti_2O_7$ was studied and a microscopic understanding of the inner workings of this material were developed. In conclusion we observe that virtually all the elements of $S_\Phi(\omega, T)$ predicted for a magnetic monopole plasma, including the existence of intense magnetization noise and its characteristic frequency and temperature dependence, are detected. Moreover, comparisons of simulated and measured correlation functions $C_\Phi(t)$ of the magnetic-flux noise $\Phi(t)$ imply that the motion of magnetic charges is strongly correlated. At the end of the chapter, directions for some future experiments are presented.

We introduced a novel SQUID-based spin noise spectroscopy technique to studies of lanthanide-pyrochlores, namely $Dy_2Ti_2O_7$ which is a Dipolar Spin Ice material. Theoretical predictions for the magnetic-flux signature of a plasma of magnetic charges $\pm m_*$ spin ice are tested for $Dy_2Ti_2O_7$. Monte-Carlo simulations predict the existence of a strong magnetization noise intrinsic to spin ice materials. Our work reports the observation of the predicted magnetic-flux noise for the first time [2]. The frequency and temperature dependence of the magnetic-flux noise spectrum $S_\Phi(\omega, T)$ predicted for $\pm m_*$ magnetic charges undergoing thermal generation and recombination (Figs. 4.2, 4.3, 4.4) is confirmed directly and in detail (Fig. 5.1). The expected transition from a plateau of constant magnetic-flux noise [3, 4] for $\omega\tau(T) \ll 1$, due to random-fluctuations, to a power-law falloff [4] for $\omega\tau(T) \gg 1$, is observed throughout (Fig. 8.1). The prediction of $S_\Phi(0, T)$ proportional to $\tau(T)$ holds true for both MC simulations of DSIM for DTO and our experiment (Fig. 8.1bottom)

These $S_\Phi(\omega, T)$ characteristics are distinctive to spin ices. This is because the magnetization-noise spectral density signature of a ferromagnet [5], a classic spin

© Springer Nature Switzerland AG 2021
R. Dusad, *Magnetic Monopole Noise*, Springer Theses,
https://doi.org/10.1007/978-3-030-58193-0_8

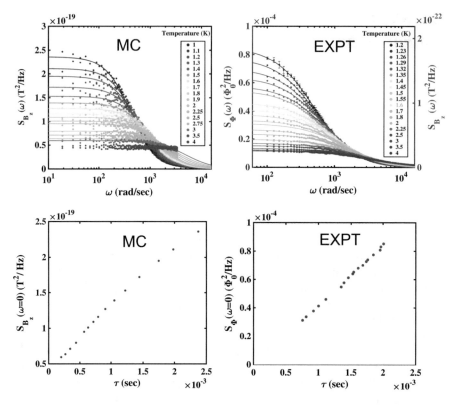

Fig. 8.1 A comparison of MC predicted $S_{B_z}(\omega, T)$ characteristics with experimental measurements of $S_\Phi(\omega, T)$ is shown here; both exhibiting equivalent characteristics to magnetic monopole generation and recombination noise predicted in Sect. 4.3

glass [6] both exhibit $1/\omega$ dependence which is very different from frequency dependence of flux noise measured in our experiment. On the other hand, the observed phenomenology of $S_\Phi(\omega, T)$ in $Dy_2Ti_2O_7$ is quite analogous to that of voltage-noise spectral destiny from GR of electron-hole pairs in semiconductors [7–9].

It is important to note that the spin ice MC calculations do not assume the existence of magnetic monopoles in spin ices, but instead find that monopoles are generated spontaneously. These calculations employ the DSIM to simulate different spin configurations of a tiny sample of $Dy_2Ti_2O_7$ at different temperatures. The fact that the spin noise calculated for a MC sample of $Dy_2Ti_2O_7$ exhibit the characteristics described by generation recombination theory for magnetic monopoles provides strong indication for the existence of monopoles in $Dy_2Ti_2O_7$. Experimental measurements of equilibrium flux noise originating from thermal fluctuations generating spin flips display a good correspondence with characteristics predicted by generation-recombination statistics of magnetic monopoles. A collective picture of both experimental observations of $S_\Phi(\omega, T)$ and MC calculations of

$S_{B_z}(\omega, T)$ from $Dy_2Ti_2O_7$ showing the magnetic monopole GR attributes (Fig. 8.1)
add to the growing evidence for the existence of magnetic monopoles in spin ices
[10–12].

The additional refinement of the generation-recombination for magnetic monopoles
in spin ices observed in the modified power law for ω is understood by studying
different magnetic charge dynamic models. The comparisons of power law 'b'
observed in experiment and the different models indicate the existence of correlations in magnetic monopole dynamics. The presentation of autocorrelation functions
$C_{B_z}(t)/C_{B_z}(0)$ show us good correspondence between experimental measurements
and the DSIM which includes both dipolar interactions and Dirac string constraints.
These observations further the understanding of the affect of correlations in spin ice
magnetic monopole plasma. Overall, we find detailed and comprehensive agreement
between current theories for thermal generation and recombination of a correlated
$\pm m_*$ magnetic monopole plasma and the phenomenology of magnetic-flux noise
spectral density in $Dy_2Ti_2O_7$ that is revealed by spin-noise spectroscopy.

8.1   Future Directions

8.1.1   Telegraph Noise

Although the monopole noise spectral density has not yet been measured in the
sub-kelvin temperature range, it is our immediate objective, but one that will
require construction of a next-generation spin noise spectrometer based on a dilution
refrigerator. We have chosen this temperature range carefully since the density
of monopoles is optimal for measurement of this kind of noise. Furthermore, we
actually anticipate that the simple magnetic-flux noise spectrum of a monopole
plasma studied here, will disappear quickly and transform into a telegraph noise
spectrum at slightly lower temperatures as predicted by generalizations of the GR
model [8]. These are challenges that can only be addressed in future research.

8.1.2   Understanding Correlations Analytically

Noting that the power law for $\omega\tau$ in the expression for spin noise does not equal
two as expected suggests a deeper look into the derivation of monopole number
fluctuations causing generation recombination noise. The last term in the Langevin
equation describes an uncorrelated stimulus to the monopole number changing as a
function of time. However it is known that the monopoles are allowed to move only
on the centers of tetrahedra the Dy spins sit on. This constrains the motion of these
monopoles in a way that is quite different from that of a free monopole plasma.
Correlations arising from such constraints are expected to be added to the last term
in equation.

8.1.3 *Magnetic Fingerprints*

Employing the technique of spin noise spectroscopy has revealed fingerprints of magnetic monopole plasma in $Dy_2Ti_2O_7$. This technology shows promise to the detection of other exotic magnetic states such as spin liquids [13], fast monopoles [14], quantum spin ices [15] etc.

References

1. A.A. Penzias, R.W. Wilson, A measurement of excess antenna temperature at 4080 mc/s. Astrophys. J. **142**, 419 (1965)
2. R. Dusad et al., Magnetic monopole noise. Nature **571**, 234–239 (2019)
3. M. Ryzhkin, A. Klyuev, A. Yakimov, Statistics of fluctuations of magnetic monopole concentration in spin ice. Fluctuation Noise Lett. **16**, 1750035 (2017)
4. A. Yacoby, N. Yao, F.K.K. Kirschner, F. Flicker, S.J. Blundell, Proposal for the detection of magnetic monopoles in spin ice via nanoscale magnetometry. Phys. Rev. B **97**, 140402 (2018)
5. M. Cerdonio, A. Maraner, S. Vitale, A. Cavalleri, G.A. Prodi, Thermal equilibrium noise with 1/f spectrum in a ferromagnetic alloy: Anomalous temperature dependence. J. Appl. Phys. **76**, 6332 (1998)
6. A.P. Malozemoff, M.B. Ketchen, W. Reim, R.H. Koch, H. Maletta, Magnetic equilibrium noise in spin-glasses: $Eu_{0.4}sr_{0.6}s$. Phys. Rev. Lett. **57**, 905 (1986)
7. R.E. Burgess, The statistics of charge carrier fluctuations in semiconductors. Proc. Phys. Soc. B **69**, 1020 (1956)
8. L. Reggiani, V. Mitin, L. Varani, *Noise and Fluctuations Control in Electronic Devices, Chapter 2* (American Scientific Publishers, 2002)
9. K.M. van Vliet, J.R. Fassett, *Fluctuation Phenomena in Solids, edited by R. E. Burgess* (Academic Press, New York, 1965)
10. P. Holdsworth, L. Jaubert, Magnetic monopole dynamics in spin ice. J. Phys. Condens. Matter **23**, 164222 (2011)
11. R. Moessner, C. Castelnovo, S. Sondhi, Spin ice, fractionalization, and topological order. Annu. Rev. Condens. Matter Phys. **3**, 35 (2012)
12. V. Kaiser et al., Emergent electrochemistry in spin ice: Debye-hückel theory and beyond. Phys. Rev. B **98**, 144413 (2018)
13. L. Balents, Spin liquids in frustrated magnets. Nature **464**, 199 (2010)
14. S. Gao et al., Dipolar spin ice states with a fast monopole hopping rate in cder2x4 (x=se, s). Phys. Rev. Lett. **120**, 137201 (2018)
15. K. Kimura et al., Quantum fluctuations in spin-ice-like pr2zr2o7. Nature Communications **4**, 1934 (2013)

Appendix A
Structure of Raw Data

A.1 Background Noise

The noise spectral density floor measured for empty pickup coil in the SNS, i.e. background noise S_Φ^{bcg}, is fit to a smooth polynomial function S_Φ^f. Since the noise floor does not vary with temperature, the same function S_Φ^f is then subtracted from all datasets $S_\Phi^{DTO}(\omega, T)$, to obtain the uncalibrated noise spectral density $S_{\Phi uncal}(\omega, T)$ dataset. An example of the raw noise spectral density, and post-processed (PP) noise spectral density is shown in Fig. A.1.

Fig. A.1 A typical noise spectrum from DTO (here at 4.00K) is shown for raw signal (green open diamond) and with smooth function fit (dark red line) to background noise spectrum (open circle, red) subtracted from the raw signal (green full diamond)

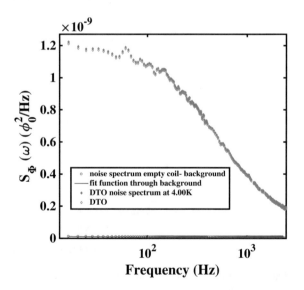

A.2   Mechanical Noise Peaks

We observed that there were some peaks in our noise spectra that were unaccounted for (Fig. A.2). The height of these peaks varied with temperature. We attributed those to mechanical noise from wires in the spectrometer (though glued down) vibrating at those frequencies. These noise peaks were manually deleted from the spectra we report in our experiment.

A.3   Fits of GR Spectral Density to Experimental Data

This post processed data is then fit to empirical equation 4.10 using Least Squares method for a BW of 16Hz–2.5kHz for all temperatures. While the plateau in flux-noise spectral density $S_\Phi(\omega, T)$ goes down to at least 1Hz for all temperatures, to optimize data acquisition times to $\sim$1 hour per temperature, for all spectra reported in Fig. 5.1, the lower limit for BW of data for regression analysis is set at 16Hz. The time constant $\tau(T)$, power law for frequency $b(T)$ and $S_\Phi(0, T)$ are free parameters in the fitting procedure and fits for all temperatures are of high quality with $R^2 > 0.99$. The residuals for these fits are shown in Fig. A.3.

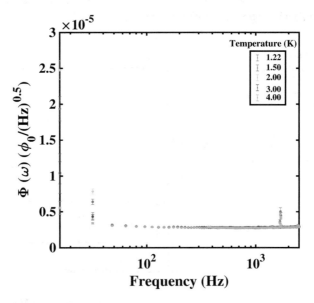

Fig. A.2 Noise floor at different temperatures exhibiting mechanical noise peaks that were removed from $S_\Phi(\omega, T)$

Fig. A.3 Residuals $S_\Phi(\omega, T) - S_{FIT}(\omega, T)$ for fits of measured flux noise spectral density (Fig. 5.1) to Eq. (4.10) are shown here for four temperatures. Here red points are for T=1.2K, yellow T=2.0K, green T=3K, blue T=4K

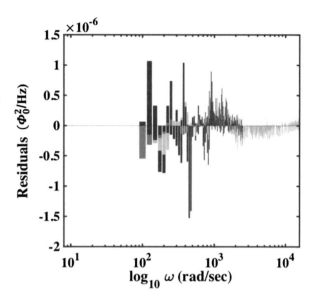

Appendix B
Dy$_2$Ti$_2$O$_7$ Flux Noise Features

B.1 Signal Strength

A typical magnetic-flux noise spectral density from a sample compared to the noise spectral density of an empty pickup coil is shown in Fig. B.1 for a bandwidth: 1Hz to 2.5kHz. The plateau of flux-noise spectral density from Dy$_2$Ti$_2$O$_7$ sits a factor of 1.5×10^6 higher than the noise floor level.

Fig. B.1 A typical spectrum of magnetic-flux noise spectral density detected from a sample of Dy$_2$Ti$_2$O$_7$ (at 1.22K) compared to flux-noise spectral density of empty pickup coil corresponding to $\sim 16.8 \times 10^{-12} \phi_0^2 / \sqrt{Hz}$. The black data points are only meaningfully visible because of a slight vertical shift in the position of 0

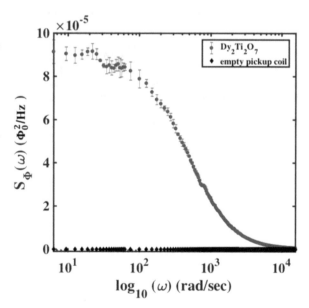

B.2   Repeatability

We established that the flux noise spectral density coming from Dy$_2$Ti$_2$O$_7$ is repeatable in different single crystals of the material as shown, for a typical example, in Fig. B.2.

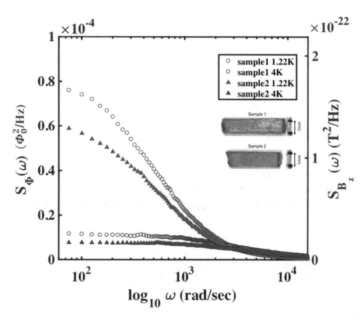

Fig. B.2 Plot of magnetic flux noise $S_\Phi(\omega, T)$ from two different Dy$_2$Ti$_2$O$_7$ rod shaped samples. We observe that the flux noise spectral density from two separate Dy$_2$Ti$_2$O$_7$ samples is very similar and therefore this experiment is quite repeatable on single crystals of Dy$_2$Ti$_2$O$_7$. The differences in magnitude and time constant are due to the geometrical differences between the two samples

Appendix C
Calibration

C.1 Inter-Calibration of Time Scales

To obtain a valid correspondence between MC step time and actual time, we assume that the MC temperature is equal to temperature of the experiment in the range 1.2K to 2K. We then plot the of generation-recombination time constant obtained from fitting $S(\omega, T)$ to $\tau(T)(\omega = 0, T)/(1 + (\omega\tau)^b))$ for both MC DSIM $\tau_{MCDSIM}(T)$ and experiment $\tau_{experiment}(T)$ respectively with temperature T as the implicit variable. We fit a linear curve to the plot with intercept =0 (Fig. C.1). The slope of the linear fit gives us the correspondence between MC step and actual time: 1MC-step = 83±11 microseconds.

C.2 Calibration of Sensitivity

The transfer function C between pickup coil and SQUID is calibrated by driving a small known flux $\Phi_{TEST}(\phi_0)$ via a drive coil (inserted into the pickup coil) through the pickup coil, and recording the corresponding SQUID output voltage VS. In this case (Fig. C.2)

$$C = \frac{V_S}{0.684} \frac{1}{\Phi_{TEST}(\phi_0)} \tag{C.1}$$

We find that C=0.015. The spectral density of magnetic-flux noise within the sample is obtained

$$S_\Phi(\omega, T) = S_v(\omega, T)/(C^2). \tag{C.2}$$

R. Dusad, *Magnetic Monopole Noise*, Springer Theses,
https://doi.org/10.1007/978-3-030-58193-0

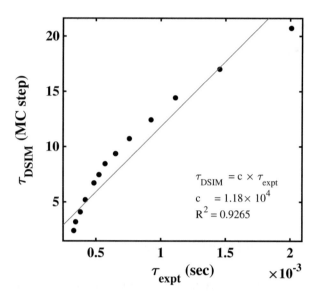

Fig. C.1 Plot of generation recombination time constant (in sec) from fits to experiment $\tau_{experiment}(T)$ vs time constant from fits to MC $\tau_{MCDSIM}(T)$ in MC step with T as the implicit variable

Fig. C.2 Here we show linear relationship between flux applied to the pickup coil via a drive coil and flux output by the SQUID. The slope between the two gives us the transfer function between the pickup coil and the SQUID

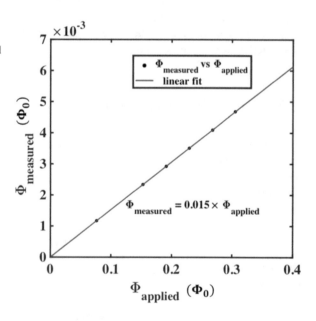

To relate $S_\Phi(\omega, T)$ to the magnetic field noise spectral density generated within our sample $S_{B_z}(\omega, T)$, we consider the cross-sectional area of the sample $\sigma = 1.4 \times 10^{-6} m^2 \pm 17\%$ yielding

$$S_{B_z}(\omega, T) = S_\Phi(\omega, T)/\sigma^2 [T^2 s] \tag{C.3}$$

Biographical Sketch

Ritika Dusad received her Ph.D. from Cornell University under the supervision of Prof. Sèamus Davis. In her dissertation work, she designed and built a novel spin noise spectrometer that has led to the discovery of Magnetic Monopole Noise in Dysprosium Titanate. The results of that study are presented in this thesis. Remarkably, the motion of magnetic monopoles and frustration of the magnets she studied can actually be heard at temperatures between 1K and 4K. Ritika Dusad was born and raised in Delhi, India and studied physics at University of California, Los Angeles as an undergraduate student. She is an expert in conceptualizing, designing and building experiments for low temperature physics. In her spare time she likes to sing Hindustani Classical Music, paint, and read fiction written by Indian authors.

Posters

1. "Spin Noise Spectroscopy of Dysprosium Titanate", Theoretical and Experimental Magnetism Meeting, Abingdon U.K., 2018
2. "Comparison of the Supercooled Spin Liquid States in the Pyrochlore Magnets $Dy_2Ti_2O_7$ and $Ho_2Ti_2O_7$", APS March Meeting, Baltimore, 2016
3. "Supercooled Spin Liquids: Status quo of Magnetically Frustrated Spin Ices", CIFAR Quantum Materials Spring Meeting, Toronto, 2016
4. "Magnetic Anisotropy Modulation in Magnetostrictive $TbFe_2$ Induced by Pressure", Maglab Theory Winter School on Frustrated Magnetism, Tallahassee, 2015
5. "Electric field induced resonant switching of domain wall chirality in nanoscale Magnetic Tunnel Junctions", Magnetism and Magnetic Materials and IEEE International Magnetics Conference San Francisco, 2013.

© Springer Nature Switzerland AG 2021
R. Dusad, *Magnetic Monopole Noise*, Springer Theses,
https://doi.org/10.1007/978-3-030-58193-0

Invited Talks

1. "Magnetic Dynamics Studies in Spin Ice $Dy_2Ti_2O_7$", Seminar, Indian Institute of Technology, Kanpur India, 2015
2. "Magnetic Monopole Noise", Seminar, University of Illinois, Urbana Champaign, 2018
3. "Magnetic Monopole Noise in $Dy_2Ti_2O_7$", Seminar, Harvard University, Boston, 2018
4. "Magnetic Monopole Noise", Seminar, Indian Institute of Technology, Powaii India, 2019

Publications

1. R. Dusad, F. K. K. Kirschner, J. C. Hoke, F. Flicker, B. Roberts, A. Eyal, G. M. Luke, S. J. Blundell and J. C. S. Davis *"Magnetic Monopole Noise"*, **Nature** 571, 234–239 (2019)
2. A. Eyvazov, R. Dusad, T. J. S. Munsie, H. A. Dabkowska, G. M. Luke, E. R. Kassner, J. C. S. Davis, A. Eyal *"Common Glass-Forming Spin-Liquid State in the Pyrochlore Magnets $Dy_2Ti_2O_7$ and $Ho_2Ti_2O_7$"*, **Phys. Rev. B** 98 214430 (2018)
3. P. Upadhyaya, R. Dusad, S. Hoffman, Y. Tserkovnyak, J. G. Alzate, P. K. Amiri, K. L. Wang *"Electric field induced domain-wall dynamics: Depinning and chirality switching"*, **Phys. Rev. B** 88 224422 (2013)
4. R. Dusad and G. Travish *"Use of Re-Acceleration and Tapering in High Gain FELs to Enhance Power and Energy Extraction"*, **Proc. of FEL Conference** Shanghai, China (2011)
5. J. McNeur, E. Arab, R. Dusad, Z. Hoyer, J.B. Rosenzweig, G. Travish, N. Vartanian, J. Xu, R.B. Yoder *"A tapered dielectric structure for laser acceleration at low energy"*, **Proc. of IPAC** (2010)

Printed in the United States
by Baker & Taylor Publisher Services